だから
(香港のアパートで)
犬は飼えないの！

リンゼイ美恵子 著
ココ・リクター 英語監修

That's Why You Can't Have a Dog in Hong Kong: A Memoir

By Mieko Lindsay
English Supervising Editor Coco Richter

まえ書き

「だから犬は飼えないの！」は香港の地元紙「香港ポスト」に2004年1月から2005年7月にかけて連載され、大幅に加筆修正及び英語版を付して2016年1月にキンドルからオンライン出版したものです。当時お世話になった杉本編集長、そして、こんな私に付き合ってくれた家族とイヌ、また英語版の監修を快く引き受けてくれたココ・リクターにお礼を申し上げます。さらに、このたびあるまじろ書房から出版するに当たり、章ごとの経費を日本円で記すなど、いくつかの改善を行いました。山田社長はじめ、多くの方々のご支援を受けることができた幸運に、心から感謝いたします。

Acknowledgment

"That's why you can't have a dog in Hong Kong" was published in the local Japanese newspaper "*Hong Kong Post*" from 2004 to 2005. I would like to acknowledge Ms. Sugimoto, Editor in Chief of *Hong Kong Post*, my family, and Coco as an English Supervising Editor. Without their help, I could not publish this renewed version of my memoir of Inu from the Kindle on-line. I also appreciate the support of Mr. Yamada of Arumajiro-shobo in publishing the revised copy of my story.

目次

まえ書き…………002

1．それなのに、出会ってしまった…………008

2．イヌと呼ばれる犬…………014

3．近所の獣医に駆け込む…………020

4．赤ちゃん用ドアを取りつけた…………026

5．マミーとダディー…………032

6．予防接種とマイクロチップ…………038

7．出費はかさむ…………044

8．クレイジーなティーンエージャーとしつけ…………050

9．犬の世話あれこれ…………056

10．案外かかる食事とおやつ…………062

11．かわいそうだけど、去勢…………068

12．しつけられるのは人間…………074

13．犬の行けるところ、行けないところ…………080

14．親友フェイジャとマズル事件…………086

15．香港を震えさせた犬の毒殺事件…………092

16．トイレは完璧になったけど…………098

17．暴れん坊の親の気分？…………104

18．留守にしてごめんね…………110

19．遊びとお気に入りおもちゃ…………116

20．「だから飼えない」VS「それでも欲しい」…………122

あと書き…………128

Contents

Acknowledgment ············003

Chapter 1 - Falling in love············009

Chapter 2 - Inu comes home············015

Chapter 3 - The pet shop lied············021

Chapter 4 - Discipline············027

Chapter 5 - Mommy and Daddy············033

Chapter 6 - Inu's debut walk············039

Chapter 7 - Tired of spending and cleaning············045

Chapter 8 - Never smack Inu············051

Chapter 9 - Helper helps but puppy is a puppy············057

Chapter 10 - Dog food's expensive! ············063

Chapter 11 - Neutering Inu············069

Chapter 12 - Disciplining people rather than the dog············075

Chapter 13 - Some places dog can't go ············081

Chapter 14 - Feijoa and the muzzle ············087

Chapter 15 - Poison shakes the community ············093

Chapter 16 - Although his toileting became perfect...············099

Chapter 17 - Parents of a rough child?!············105

Chapter 18 - Family should always be together············111

Chapter 19 - Japanese pet shops are interesting············117

Chapter 20 - "That's why you can't!" vs. "Still I want one!" ············123

Postface············129

香港 この本の舞台
― Hong Kong ―

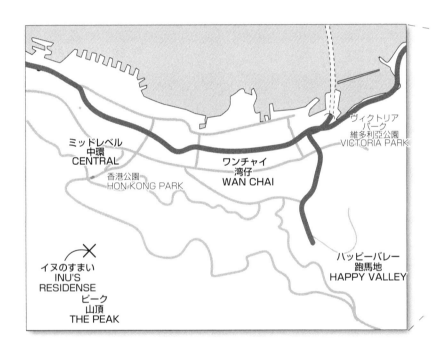

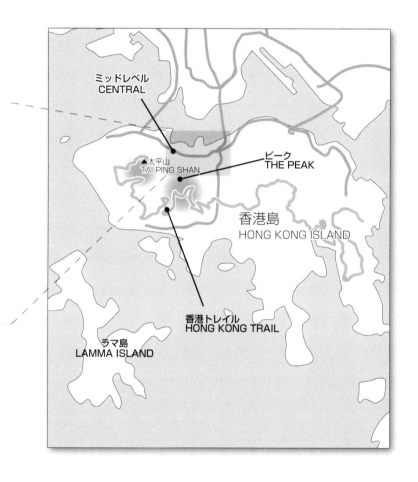

１．それなのに、出会ってしまった

　「犬、飼ってえ」とお子さんに泣きつかれている人は多いと思う。一人暮らしや夫婦者でも、大都会香港の生活に疲れ、日本を離れた寂しさに、「犬か猫でもいたらなあ」と考えがよぎるのではないだろうか。

　でも、そうやすやすと飼わないのは「香港のマンション住まいで犬を飼うなんて、お金を払って苦労を背負い込むだけ」というのがいかに正論か、わかっているからだと思う。

　そのことはよおくわかっていたはずなのに……。なぜか飼ってしまったのが私たちだ。ええい、こうなったら、どんな苦労を背負い込んだのか、皆様にとくとお見せしようではないか。これを読んでお子さんたちに「だから犬は飼えないの」と説得していただきたい。そして言葉の通じない生き物相手に、必死に社会生活をさせようとする私たちのこっけい振りを、笑っていただきたい。

　でも、でも、もしかしたら、小動物のもつヒーリングパワーに軍配が上がり、これを読んでさらに犬が欲しくなる人も、いるかもしれないけれど……。

　時は 2001 年。私たち、子なし夫婦にとって、香港は「仮り」かつ「借り」の住まいであり、できるだけ「物は持たない増やさない」の精神で生きてきた。ああ、それなのにそれなのに……。

　５月のある日、ハッピーバレーを歩いていた私は、あるペットショップの前で一匹の子犬に釘付けになってしまった。

　その夜、夕飯の席で夫に「犬、飼いたい」と切り出すと、夫は絶句。だが、実は夫もひそかに飼いたかったらしく、翌日にはワンチャイの SPCA（動物愛護協会）へ出かけた。

Chapter 1 – Falling in love

Many parents have suffered from their children begging them to have a dog. Well, it's not only children who may cry for the company of a dog. When you feel lonely living away from your hometown, or perhaps if you are single or a couple without children, you might consider having a dog to share your life. But don't be rash. Making a home for a dog in a high-rise apartment in Hong Kong is a lot of trouble, as my husband and I learned when we decided to get a puppy. I'm not even sure why we made this decision but we did, and now I want to share with you all our troubles and struggles.

You may recognize us. Perhaps you have seen us on the streets trying to discipline our dog to be a good member of the community and society. An absurd sight, I know. You may have laughed at us, telling your children, "Look at those crazy people. They are in deep trouble. That's why you can't have a dog in Hong Kong." Or you may decide you still want one, and just go for it, like we did. But don't say I didn't warn you.

That was in 2001. For most childless couples like us who are foreigners and thinking of living only for a short while in Hong Kong, it is a good idea not to fill your life with too many things. I followed this rule very well until one fateful day in May. I was walking along the street in Happy Valley when I caught sight of a puppy sitting in a cage at the pet shop. I stopped and stared, unable to move, so taken was I by the sight of this cute, furry

でも、縁がなかった。ほとんどの犬がご予約済み。その上驚いたことに、私と夫の間で意見が合わないのだ。夫が好もしく思う犬は、むさくるしくて横柄な態度に見える。結婚して10年近いが、相手の犬の趣味に、お互い改めて驚かされた。

仕方なくSPCAを出ると、「きのうの犬に一目会いたい」という気持ちで、ハッピーバレーへ歩いた。ああいう時の気持ちは不思議だ。「あの犬を見ないでは死んでしまう」と思えてくるのだから。一説によると犬が人を選ぶというから、あの犬が私にテレパシーを送っていたのかも。

ペットショップで檻から出した子犬を抱きしめた夫も、手放せない様子。「いくら?」と聞くと、およそペットショップにふさわしくない、綺麗な爪をしたサイケデリックな女性店員が「5,800」とのたまう。

「値札は3,800ドルじゃない」と抗議したが、「それは間違い。血統書付きだから」と軽くあしらわれた。

子犬はオスで、一応ゴールデンレトリーバー。一応というのは香港で代々掛け合わされてきた種で、元祖イギリスのゴールデンに近いが、香港のお金持ちが連れているアメリカ産やオーストラリア産に比べて、色が薄くて細身で毛が貧弱。でも、子犬と恋に落ちている私たち夫婦には、そんなことは見えなくなっている。

私たちはその犬にお買い上げ予約を入れると、「きれいにしておくから明日取りに来て」と言うサイケな店員に従って、家路に着いた。かわいさには惚れたが、大型犬を飼う自信のなかった私は、その夜、巨大化した犬が我が家をトイレにしてしまう悪夢にうなされた。後から考えるとなんとも示唆的な夢だった。

little creature.

At dinner that night, I told my husband, hesitating just a bit, that I wanted a dog. He didn't respond but the next morning, he suggested we go to the SPCA - that's the Society for the Prevention of Cruelty to Animals - to choose one. I guess he wanted a dog too.

We had no luck at the SPCA because most of the dogs there had already found homes and were just waiting to be collected by their new families. But we learned, my husband and I, that our taste in dogs was totally different. Having been married for nearly ten years, this was a surprise. The dog he thought was so cute – a scruffy looking, wirehaired species with an arrogant expression - was not my type. I wanted one of those golden haired dogs that always looked so happy.

After we left the SPCA (feeling rather disappointed), we walked down to Happy Valley, hoping to see the puppy that had been in the window the day before. My mind was consumed with thoughts of that puppy, so much so that I had the feeling that if I failed to see him, I would have died. Some people say it is the dog that chooses the master. I think that may be true because that puppy was sending me a strong wave of telepathy that I had to take him home.

Fortunately, the puppy was still there at the shop, and once my husband and I held him in our arms, we had to have him. We asked the shopkeeper for the price of the puppy. She was a young woman dressed in a psychedelic outfit with brightly colored nails.

子犬の目はフラッシュで青く光る

Puppy's eyes got blue with the reflection of flush rights.

```
ここまでの経費：
(香港ドルは当時1ドル約16円、または約0.13米ドル)
繰越し金：                    ￥0
SPCA入会費：              ￥1,600
SPCA寄付金：              ￥6,400
..................................................
合計：                    ￥8,000
                          チーン！
```

"$5,800 Hong Kong Dollars," she said.

The sign said $3,800 but she said it was wrong because the puppy had a pedigree.

The puppy was a boy and he looked similar to a golden retriever but he was thinner and had paler, shorter hair than many of the other pedigreed golden retrievers that were imported from the US or Australia by wealthy families in Hong Kong. The puppy was a Hong Kong breed, she said, and closer to the original English breed. She gave us a hard sell but we didn't really care because we had already fallen in love with him. We paid a deposit for the puppy and the psychedelic woman told us to pick him up the next day so she could get him nicely groomed for us.

When we got home, I became nervous, wondering if I would be able to take good care of him and how big he would become. That night, I had a terrible dream that he became huge and treated all parts of our apartment as his toilet. I know now that this was a premonition of our future life with furry company but still, we had to have that dog.

Expenditures
(Currency rate of HK$1 in 2001 was about 16yen or US$0.13)

Balance carried forward :	HK$ 0
SPCA joining fee :	HK$100
SPCA donation :	HK$400
Total :	HK$500
	Ka-ching!

2．イヌと呼ばれる犬

その日は日曜日だった。気持ちとしては、朝からペットショップに走っていって、前日予約を入れた子犬に会いたいほどだ。が、サイケな店員に「4時に引き取りに来て」と指定されている。それまで友人夫婦と食事などして過ごす。話題はもっぱら「犬について」。聞かされるほうは迷惑かもしれないが、ほかの事は頭に入らない状態なので、しかたない。

4時ぴったりにペットショップに着くと、そこには綺麗にグルーミングされた私たちの子犬が待っていた。もう手放せないとばかりに抱っこしたまま、必要なものを買い揃える。

まず、大型犬用の檻。折りたたむと10センチぐらいの幅になるが、結構重い。他にドッグフード。子犬用引き綱と首輪。甘やかしてはいけないと、おもちゃは2個だけにして、あとはブラシ。

さて、大量の買い物も終わったのに、4時に約束した獣医が来ない。サイケなファッションで決めたペットショップの女店員は、電話で催促している。が、緊急の患者だと言って現れない。待つこと2時間以上。

その間に、犬が咳を始めた。子犬の選別として正しいとは思えないが、私は兄弟2匹で檻に入った中からおとなしいほうを選んでいた。もともと大型犬、特にこのゴールデンレトリーバーは吠えない種だ。私の腕の中で愛らしい笑顔（に見える）の子犬は、毛こそふわふわだが、からだはがりがりに痩せている。細い肩を小刻みに震わせながら咳をする様子は、はかなさを漂わせている。

「この咳大丈夫かな？」と質問する私と夫に、サイケな店員は「これは咳じゃない。水飲んでむせてるの」という。檻に返すと横に

Chapter 2 – Inu comes home

The next morning, we wanted to dash to the pet shop first thing to pick up our puppy but the psychedelic lady told us we couldn't come until 4:00 pm, so we met some friends for lunch. We couldn't talk about anything but "our dog". Our friends didn't seem very interested but we couldn't help ourselves.

At 4:00 pm sharp, we arrived at the pet shop, overwhelmed to see our puppy looking nicely groomed and waiting for us. He was so cute we couldn't put him down, so we held him in our arms as we shopped for his needs.

We picked the cage for a large size dog, which was quite heavy but could be folded down into 10cm thickness. We bought dog food, a collar and leash, two toys (we didn't want to spoil him too much on the first day), and a grooming brush. Somehow we'd managed to fill the shopping cart.

We were ready to leave but the veterinarian who was supposed to provide the health check to release the puppy hadn't arrived yet. The psychedelic lady called the vet many times, telling us that she was in surgery. We waited for more than two hours.

It was during this period – after the shopping and before the vet arrived – that the puppy started to cough. I noticed our dog was quieter than the other dog from the same cage and wondered if I'd chosen the wrong one. When my puppy coughed, his thin little body shuddered. I feared this fragile life might end soon.

We asked psychedelic lady if the cough was caused by any

なる様子は、ぐったりしているようにも見えるが「子犬はよく眠るから」と説明をうけ、目を見合わせる私と夫。「どうしようか」と問う私に「じゃ、違うのをもらう？　ペットショップは構わないって言うよ。だけど僕は駄目だ。もうこの犬でなくちゃ駄目なんだ。そういうこと分かって商売しているんだよ」と、納得のお答え。私もこの犬でなくっちゃもう駄目だ。他は考えられない。

　３時間待たせた獣医は犬が健康体である証明書を出し、耳の洗浄方法や予防注射の予定を説明した。子犬は３ヶ月といわれたのに、獣医は「もう少し小さい。２ヶ月半かな」という。診察の間、犬は咳をしなかったが、獣医はそそくさと帰っていった。

　やっと愛しの犬を手にいれ、大量の犬用品をぶら下げて、私たちはタクシーに乗った。

　途中で夫が「犬は日本語でなんていうの？」と問うので「イヌ」と答えると、それを名前にしようという。日本語を話せないアメリカ人の夫の日本語語彙が増えるから、変だけれど「ま、いいか」と賛成する私。

　命名された当のイヌは、夜になると咳が止まらない。「あした朝一で、他の獣医に行かなきゃね」と、不安な第一夜を過ごしたのだった。

disease.

"That's not coughing," she said. "Puppy, he just choke on the water."

Puppy was getting heavy in our arms so we put him back in the cage. He lay down immediately as though he didn't feel well.

"Puppies sleep all the time," the lady said.

My husband and I looked at each other. "What do you think?" I asked him.

"Do you want the other one?" my husband asked, pointing to the frisky one now wagging his tail at us. "The shop doesn't care, but I do care."

It may have been crazy but I felt the same. I couldn't take any other dog than the one I'd first laid eyes on. Despite the cough and his rather sickly look, I had to have this puppy.

The vet finally arrived, but it was another hour before she issued us the health certificate. She explained how to clean puppy's ears and the schedule for his vaccinations. And though the shopkeeper had told us the puppy was 3-months old, the vet said he was younger than that, probably by two weeks or so,

Finally, we had our puppy and we moved to the street together with all of our shopping items, and hailed a taxi. I didn't know that you could get in a taxi with a dog in Hong Kong. How convenient was that!

My husband, who is American and hardly speaks any Japanese, asked me, "How do you say "dog" in Japanese?" "Inu," I said.

He suggested that could be the name for our new puppy. Although it's strange to name a dog a dog, I thought it would

17

```
ここまでの経費：
繰越し金：          ￥8,000
イヌ：             ￥92,800
大型犬用檻：        ￥15,680
子犬用ドッグフード：   ￥4,160
引き綱とチェーン：    ￥4,480
おもちゃ：            ￥400
耳洗浄液：          ￥1,920
ブラシ：            ￥1,600
....................................................
合計：           ￥129,040
                  チーン！
```

手のひらに乗りそうなほど小さかったイヌ
Inu was so small. We felt he could sit on our palm.

expand my husband's Japanese vocabulary so I said okay.

Inu started coughing in the evening. We spent the night worrying for Inu whose cough seemed to be getting worse.

Expenditures :

Balance carried forward :	HK$500
The puppy (Inu) :	HK$5,800
Large size cage :	HK$980
Dog food:	HK$260
Leash and chain :	HK$280
Toys :	HK$25
Ear cleaner :	HK$120
Brush :	HK$100

Total : HK$8,065

Ka-ching!

3．近所の獣医に駆け込む

　一目で恋に落ち、良くない買い方だと知りながらペットショップから譲り受けたゴールデンレトリーバーの子犬。「イヌ」と命名されたその犬は、引き取った初日から咳が止まらない。「ペットショップが用意した獣医なんて信用できない」とばかりに、翌朝、私と夫はミッドレベルにある近所の獣医に連れて行った。

　まだ予防接種が終わっていないので、地面や病院の床に、直接置くことはできない。がりがりの体を抱いて待っていると番が回ってきた。体重を測ってもらうと３キロ弱だ。

　診察した獣医さんはまず「耳ダニ」と告げる。良くあることらしいが、犬にとっては掻くに掻けないつらさなのだと説明される。「毎日この液で洗浄して」と見せられた薬はペットショップの獣医と同じ物。だからあんなに耳洗浄を強調したのか、と夫と目を見合わせる。それにしても「耳ダニ」ぐらいで買うのをやめたりしないから、はっきり言ってくれれば良いのに。

　そして肝心の咳のほうは「ケンネルコフというペットショップでかかる咳だと思うけど」と言葉を切り「もしかしたらジステンバーかも」と、ショックなお言葉。「この小ささだから、最悪の事態に備えるように」と私たちの覚悟を促し、抗生物質を処方してくれた。しかも「この犬が３ヶ月なんてまるで嘘。２ヶ月やっとだね」と言う。歯の成長ぐあいを見ると、専門家には一目瞭然なのだそうだ。あとから送られてきた血統書で、この言葉が正しいと分かるのはまだまだ先のことだ。「ペットショップにしてみれば、できるだけ早くに売ってしまいたい。手がかからなくなる３ヶ月で譲り受けるのが理想と言われているから、そういう嘘を

Chapter 3 – The pet shop lied

Although we knew buying a puppy from the pet shop was not the best place to go, we could not resist because we'd fallen in love with him at first sight. But Inu couldn't stop coughing from the first day he came home. We went to the vet in our neighborhood in the Mid-levels as soon as it opened the next morning. Since Inu hadn't completed his vaccinations yet, we held him in our arms in the lobby of the clinic. He felt so skinny. When our turn came, the vet put him on the scale. He weighed less than 3 kilograms.

We next learned that Inu had ear mites. It's not uncommon for dogs to have this problem but it's uncomfortable for them because they can't scratch inside their ears. The doctor showed us how to clean Inu's ear, using the same liquid given to us by the vet at the pet shop. That's why she instructed us repeatedly to clean his ears though she never mentioned the ear mites.

The cough, we learned, was likely something called "kennel cough" but the vet couldn't rule out the possibility that it was distemper. He prescribed an antibiotic but cautioned that it might not cure him.

"You must prepare yourselves for the worst since the puppy is so young," he said. "And it is a complete lie that this puppy is 3-months old. He's not even 2-months."

The doctor knew this from examining Inu's teeth.

"Pet shops want to sell puppies as soon as possible so it's quite

つく店は多い」と説明された。

　それにしても3ヶ月後に送られてきた「血統書」には犬の生年月日が入っているだけで、母親欄も父親欄も空白なのだった。どこが血統なんだか。血統にはこだわらないが、血統書付きを理由に高い値段を吹っかけられたのだから、詐欺に合ったとしか思えない。

　もっともきちんとした買い方を研究せず、「この犬が欲しい」という感情のおもむくままに買ってしまったこちらにも、手抜かりはある。それに「この犬でなければイヤだ」と思ったあの時の強い感情は、どうしようもないもので、他の犬では代用できなかったのだからしょうがない。こういうのを運命と言うんだろうか。

　さて、家に帰ると子犬に抗生物質を飲ませる戦いが始まった。もがいて嫌がる子犬を夫が押さえ、針無し注射器の中に仕込んだ薬を私が口の横から差込み、一気にイヌののど奥に押し出す。入れ方が浅いとぺっぺと吐き出してしまう。これを1日3回。夫が会社に行っている間は私が一人でやって見せねばならない。

　弱っていたせいか、人恋しさに泣くことはあまり無かったが、初日のウンチはゆるかった。環境が変わって緊張しているのだろう。などと同情したのは最初の2日のみ。抗生物質で元気を取り戻し、我が家に慣れたイヌは、夢で恐れたとおり、我が家を巨大トイレと化して飛び回り、私をてんてこ舞いに追い込むことになる。

22

common to tell a lie like that," the vet said. "It's a common knowledge that you should wait to separate puppies from their mother until they're at least 3-months old."

Several months later, when the pedigree certificate arrived, we learned our doctor was correct. And the certificate provided no information besides Inu's birthday. So what did pedigree mean? We didn't care if the dog was pedigreed or not but we paid more money because of that and didn't feel it was really worth it. Between that, the ear mites, the kennel cough and the lying about Inu's age, we felt a bit cheated.

But, to be fair, we hadn't done our research. We just followed our desire to have that puppy. I honestly never wanted something so strongly in my life, and my husband felt the same way. No other dog would do. We were uncontrollable. Inu was our destiny.

Back home, we struggled to feed Inu his medicine. He was really unhappy about having to take that. Three times a day, I put the antibiotic in the injection tube, holding him tightly as I craned open his little mouth to get the medicine in. If the injection was not deep enough in his mouth,he spat the medicine out. My husband helped me to hold Inu but while he was at the office I had to do it all by myself.

That night, he didn't bark or cry out but he seemed pretty weak. Maybe he was nervous in the new environment. But after two days, our worries quickly passed. Thanks to the antibiotic, Inu recovered quickly and got used to our house. Within a couple of days, he was racing about the apartment, making it his toilet, just as I'd seen in my nightmare.

ここまでの経費：
繰越し金：　　　　　¥129,040
診察料：　　　　　　¥4,000
薬代：　　　　　　　¥28,800
....
合計：　　　　　　　￥161,840
　　　　　　　　　　チーン！

緊張していたイヌに同情したのは最初だけ

We were sympathetic to his nervousness

only for the first couple of days

Expenditures：
Balance carried forward： HK$8,065
Medical consultation： HK$ 250
Medicine charge： HK$1,800
....
Total： HK$10,115
Ka-ching!

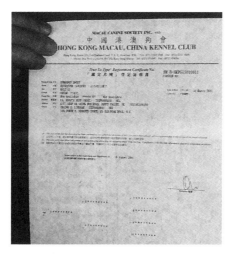

血統証は付いていた、けれど・・・

The certificate didn't seem to be certifying anything.

4．赤ちゃん用ドアを取りつけた

　一目ぼれで買ってきたゴールデンレトリーバーの子犬、「イヌ」。はかない面持ちで我が家に連れてこられ、獣医さんに「死ぬかもしれないと覚悟して」とまで言われたのに、抗生物質と私たちの愛情に満ちた（と思いたい）お世話のおかげで一週間後には元気一杯、家中をところ狭しと飛び回るようになった。

　元気になって困るのが下の世話。まだ2ヶ月の子犬は膀胱が小さくて、おしっこは日に15回以上、ウンチも5回はする。小さな水溜りに気付かず、足を踏み入れてから悲鳴をあげることもしばしばだ。掃除は徹底的に匂いまで拭き取らないと、その場所に繰り返しおしっこするようになる。

　トレーに新聞紙を敷いてイヌ用トイレを作り、おしっこの匂いが少しついた紙を残しておく。こうするとイヌが同じ場所で用を足すようになると人から聞いたが、ちゃんとできるのは10回のうち3回程度。どこにお漏らししたか見極めようと、目をらんらんと光らせ、右手に雑巾、左手に消臭スプレーを持って家中を歩き回る私は、他人が見たらかなり恐ろしかったはずだ。

　夫も私もそれぞれ「犬の飼い方、しつけ方」といった本を買ってきて、夜遅くまで読んで勉強した。二人とも、子供の頃に庭付きの家で犬を飼った経験はあるが、マンションで飼うのは初めてで、まったくの手探り状態だ。

　本によると、子犬にはあまり自由を与えすぎてはいけないらしい。「ここがあなたのテリトリー」と犬が落ち着く場所を教えなければならないとある。家中自由に走らせるなんて、子犬の間はもってのほかなのだ。

Chapter 4 - Discipline

Thanks to the antibiotic and our loving care – we're pretty sure love was part of the reason for his survival – within a week, Inu had recovered. But now we had the problem of his toileting. The bladder of a two-month old puppy is apparently quite small so Inu needed to go a lot. I'm not kidding. Fifteen times a day for pee and five times a day for poo. And even then, I kept putting my foot in his little puddles in the apartment. Cleaning became a big job. It had to be thorough to eliminate the odor otherwise he would pee on the same spot repeatedly.

We made him a toilet by putting old newspaper with a little smell of his own urine on a tray. We heard this was the recommended technique for toilet-training dogs, but it didn't seem to work for our Inu since his success rate was about three out of ten. I patrolled the house with a damp cloth in my right hand and odor elimination spray in my left hand, frantically looking in each room to be sure I didn't miss a spot.

We bought several books on "how to discipline your dog", studying them late into the night. According to the books, a puppy should not be given too much space. We needed to set a territory for him to settle; it wasn't a good idea to allow him to run around all over the house.

So we moved Inu, his toilet tray and his cage to the washing room by the back door. We probably should have just gotten rid of the cage because Inu refused to use it. What a waste that

27

そこで、イヌはキッチンの奥、洗濯部屋と呼んでいる勝手口の横に、おトイレと檻と一緒に置かれることとなった。もっともイヌは檻に全然入らず、980ドルは無駄金だったのだが。

そして、夫が会社の帰りに買ってきた、ベビー用ドアをキッチンと洗濯部屋の間、キッチンとリビングルームの間にそれぞれ取りつけた。夜寝る時は洗濯部屋、昼間はキッチンまで走ってきて良し、特別良い子の時はリビングまで出してあげる、という具合だ。だが、良い子だと思ってリビングに出すと、そこでピーとお漏らしする。まるで狙い撃ちのようだ。

「あたしもう、限界。今日は32回もイヌのおしっことウンチを拭いたの」と、ある晩夕食の席で、私は夫に泣きを入れた。「じゃ、どうする？　ペットショップに返すの？」と問う夫の言葉に、私はテーブルをドンとたたいて立ち上がり、「いいえ、あのイヌ、絶対にしつけてみせる」と宣言した。これは、本人は忘れていたが夫が覚えていたことで「あの時のきみ、結構こわかったよ」と言われた。しょうがないじゃない、こっちだって必死だったのだから。

ここまでの経費：

繰越し金：	¥161,840
ベビー用ドア2枚：	¥25,600
犬の躾け方の本二冊：	¥3,840

合計：	¥191,280
	チーン！

HK$980 was!

Armed now with the knowledge that Inu's space needed to be limited, my husband bought two baby safety doors at the children's shop, installing one between the washing room and kitchen and the other between the kitchen and living room.

"Now," we explained to Inu, "you sleep in the washing room, and you can come out to the kitchen during the daytime. And if you are very good, we will let you out into the living room."

As you might imagine, it didn't work very well. Every time we told Inu he was a good boy and let him out into the living room, he peed, just like he was taking a potshot.

"That's enough!" I cried to my husband over dinner one night. "Do you know how many times I cleaned his pee and poo today? Thirty-two times, thirty-two!"

"So what do you want to do?" my husband asked. "Return him to the pet shop?"

I stood up and banged my fist against the table. "Of course not!" I replied. "I will discipline that puppy."

My husband later told me that I was very scary when I did that. I feel a little badly about having this rather bad attitude but caring for a puppy is an exhausting business.

Expenditures :	
Balance carried forward :	HK$10,115
2 baby safety gates :	HK$1,600
Dog training books :	HK$240
Total :	HK$11,955

Ka-ching!

ドアを取りつけてイヌにテリトリーを分からせる
Installing baby doors to make Inu understand his territory.

31

5．マミーとダディー

　咳をしながら我が家に引き取られてきて、イヌと名づけられた犬が、私たちにもたらした変化は多大なものだった。1日3回の食事と耳掃除、トイレの世話、といろいろあるが、ともかく自分たちが手をかけなければ生きていかれない存在が、家で待っているというプレッシャーはそれまでの私たちにはないものだ。

　何より驚いた変化は、夫の態度だ。会社から帰ると「ハーイ、ダディーは今帰ったよ。良い子にしてたかなあ」と一通りイヌとじゃれあう。その声は、ついぞ私に向かって発せられたことのない、喜びに満ちた声だ。だいたい「ダディー」というのが夫のことだと理解するまで、私には少々時間が必要だった。それだけではない。「マミーの言うことをちゃんときいたかな？　男の子なんだからマミーを守らなくっちゃ」などと話しかける。

　ゴールデンレトリーバーはガードドッグにはならないので「マミーを守る」は完全な冗談としても、いったいいつから私はイヌの母親になったのか。

　でも、子犬の毛は本当に柔らかく、かわいい瞳や手足のしぐさを見ていると、マミーと呼ばれてもしょうがないか、という気がしてくる。それに、イヌのおかげで夫が幸せなら、ダディーと呼んでも良いじゃないかと思えてくる。会社ではそれなりのストレスだってあるだろうけど、イヌになら素直になれるみたいだから。

　イヌが大きくなってからは、散歩で知り合う犬友達がたくさんできたが、みんな飼い主のことをマミーとダディーで呼んでいる。例えばイヌの親友ブルマスティフの「テディ」の飼い主は、テディのマミーとダディーで、本名は存じ上げない。ピークでよく

Chapter 5 - Mommy and Daddy

Our little puppy drastically changed our lives. Meals three times a day, cleaning his ear to prevent mites, and toilet training. We had so many things to do for him. It was stressful living with something that couldn't survive without our care. And this was all new to us.

Among the changes that was most surprising to me was my husband's attitude. When he got home from work each day, the first thing he'd say was, "Daddy is home. Have you been a good boy?" Then he'd start playing with Inu. He used such a sweet, gentle voice, that, frankly, I can't remember him ever using with me. It took me a minute to understand who "daddy" was. And this was just the beginning of it.

"Did you listen to Mommy? You are a boy, you have to protect your Mommy, alright?" he'd say, blathering on and on.

I didn't have the heart to tell him that golden retrievers are not guard dogs so "protecting mommy" was a joke, and since when did I become the mommy of a dog?

But looking at Inu's adorable gestures and lovable paws and eyes, I didn't mind being called his mommy. And besides, my husband was happy thanks to Inu. So I decided it was okay he called himself the "daddy."

We had to wait to take him for a walk outside till he got 3-months old. And in getting acquainted with other dog owners, we found everybody called themselves "mommy" and "daddy"

出会う日本人のご夫妻も「リリー」のマミーとダディーだ。

　イヌをかわいがるあまり、夫はイヌをおなかや頭に乗せてじゃれついていたが、これはお勧めできない。当時３キロだったイヌは１歳半の成犬になると 33 キロになった。そして相変わらず同じように夫の頭やおなかに乗りたがるのだ。甘やかすから自業自得だ。

　さて、本によれば犬には常におもちゃが必要なんだそうだ。おもちゃが無ければ靴やスリッパを齧りだす。私は知らなかったが、犬は人間と同じように乳歯と永久歯が生え変わる。それまで何でもかんでも齧るのだ。一番気に入ったのは、ダイニングテーブルの下のカーペット。ループが出ていて引っかかるところが楽しくてしょうがない。噛みついては繊維を引き出して破壊した。イヌが１歳になって落ち着く頃まで待って買い換えた。テーブルや椅子の木枠も犬の歯型がばっちりだ。新聞紙をめちゃくちゃに齧り裂くのも、大好きな遊びだ。うっかりおもちゃにされた携帯電話は機能しなくなった。

　齧るのを止めようと、犬の嫌がる匂いをスプレーにした、リペラントというものを撒いてもみた。が、２日と持たない。その上、犬の嫌いな匂いは人間にとってもすごく不快なのが分かり、めったなことでは使用しなくなった。

　それまで大人二人で整然と暮らしていた私たちは、犬のかわいさにヒーリングパワーを感じながらも、破壊され、トイレ臭の染みついた我が家に、新たなるストレスを感じ始めたのだ。

of their dogs. One of Inu's best friends was Teddy, a bull mastiff, and we only knew his owners as "Teddy's mommy and daddy." We never found out their real names. Similarly, a Japanese couple we often ran into at the Peak were only known to us as "Lily's mommy and daddy." That's just how it was.

My husband loved Inu so much and played with him, pulling Inu onto his stomach and raising him above his head. I wouldn't recommend this to anybody. Back then, Inu weighed only three kilos, but at one and half years old, he weighted 33kilos. That's 70 pounds! And Inu still wanted to do the same thing, lying across my husband's stomach and getting over his head. My husband spoiled the dog, for sure.

According to the books, puppies need toys. Without toys, they chew and destroy shoes and slippers or anything they can get their mouths on. I didn't know that dogs have baby teeth that are replaced with permanent teeth like humans, and during the teething period, puppies chew anything. Inu loved the carpet under the dining table and never tired of pawing and pulling the loops from the carpet. He bit and chewed and snarled the carpet, ruining it completely. We gave up trying to stop him, deciding that we'd wait until he was a year old and then we'd buy a new carpet. Inu's teeth marked the armrest of the sofa, the legs of the dining table, and pretty much anything else in the apartment made out of wood. Chewing up newspaper was another source of fun for Inu. He even chewed my mobile phone!

In defense, we tried spraying "dog repellant" on every piece of furniture but it didn't last for more than two days; and the

ここまでの経費：
繰越し金：　　　　　¥191,280
買い換えたカーペット：¥35,200
犬よけリペラント：　　¥1,280
....
合計：　　　　　￥227,760
　　　　　　　チーン！

イヌにかまれ続けた椅子の木枠

Inu liked to chew the wooden frame of furniture.

uncomfortable smell for dogs was awful for us too, so we tried not to use it unless there was no other choice.

Before Inu came along, my husband and I had been used to a quiet, organized lifestyle. Though Inu gave us a lot of healing power radiating from his lovable cuteness, we started to feel a new type of stress and pressure from a life with a puppy. Our house was now always messy and smelled like a public toilet.

Expenditures：	
Balance carried forward：	HK$11,955
New carpet：	HK$2,200
Dog repellant spray：	HK$80
Total：	HK$14,235
	Ka-ching!

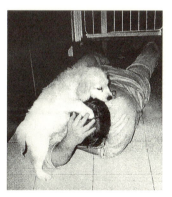

いくらかわいくても、こんな遊びはお勧めしない
Although cute, don't play like this with a puppy as he will grow big.

6．予防接種とマイクロチップ

　かわいさにほだされて衝動買いしてしまったゴールデンレトリーバーの子犬、「イヌ」。我が家に慣れて元気になると、家中おしっこをして我が物顔で走り回るようになった。その掃除も大変だったが、さらに私を困惑させたのは、「常に一緒にいたい」というイヌの態度だ。いや「いたい」などという願望ではなく「いなければならない」という強い主張なのだ。

　子犬の間は餌も1日3回あげなければならず、一人（匹）にしておくと、新聞紙や置物をめちゃめちゃにする、いわゆる「破損行為」に出るので、うかうか外出もできない。家では私にずっとついて回り、料理する間も私の足元に横たわるので、間違って踏んづけたり、包丁や熱い物を落としたらどうしようと、緊張させられる。

　さらに困ったのが、私がお手洗いに入ると、ドアの外で際限なく吠えることだ。子犬の甲高い声でワンキャンと吠え続けるのには参る。とうとう私は用を足す間、イヌをトイレに入れることにした。ドアを開けると尻尾を振って入ってきて、私の足元にうずくまり満足した様子だ。ハイハイするようになった赤ちゃんの母親が、一時も目を離せないからトイレのドアを開け放って用を足すというのは良く聞く話だが、子犬の場合も状況は似ている。

　5月の半ばに月齢約2ヶ月で我が家にきたイヌは、7月始めに予防接種をすべて終え、首の後ろに身分証明であるマイクロチップを埋めてもらい、晴れて外に出られる身となった。かわいいけれど、家の中にこもったイヌの匂いに閉口していた私たちは、本当に嬉しかった。

Chapter 6 - Inu's debut walk

Aside from the now incessant job of cleaning up after Inu, I had to get used to Inu's desire to be with me all the time. This was baffling to me. No matter what I was doing in the apartment, Inu demanded that he be allowed to come too. In fact, Inu behaved badly whenever I left him alone in the house. He'd chew up newspapers or any loose items left near the floor. This "destructive behavior" stemmed from the stress of the loneliness he felt, or so the vet told us.

So now I could hardly leave the house. And when I was home, Inu followed me everywhere. When I was cooking, he'd lay down between my legs. I was always worried I would drop my knife or hot liquid on him, or step on him by mistake. He made me so nervous.

He'd yelp if he couldn't see me. This created a troublesome situation whenever I went to the bathroom. Puppy's high-pitched yapping pierced my ears. To appease (and quiet) him, I'd give in and let him in. He wagged his tail so happily, curling up at my feet while I tried to take care of my business. We often hear that mothers with babies who start crawling can't even use the toilet without keeping their eye on the baby. With Inu, the situation was more or less the same.

It was now July and Inu had completed all his vaccinations and been tagged with a microchip in his neck so he was finally allowed to go outside. This was exciting because by then, the

それまでにイヌが受けた予防接種は混合ワクチンが３回、そして３ヶ月に成長した時点での狂犬病予防が１回だ。その後も混合ワクチンは毎年１回、狂犬病予防は３年に１回、受け続けなければならない。他に、心臓に寄生するフィラリア予防のワクチンを６ヶ月の時に受けた。お医者を嫌がるかと心配したが、イヌが気づかぬうちに、首筋にさっと注射針を入れる獣医さんのてぎわは鮮やかで、本来人なつっこいゴールデンレトリーバーは、すぐになついた。医者に行くのを喜ぶ、変な犬だ。

　さて、初めての散歩に出たイヌだが、そうそう簡単に歩くものでもない。初日は状況がよく飲み込めなかったようで、私たちの足元に絡まり、顔をうかがっていた。だが、恐がり気質ではなかったらしく、すぐに尻尾を振ってついてくるようになった。

　我が家の散歩コースはミッドレベルからピークへ続く、急勾配の道なのだが、この年は雨が多く、路上を滝の様に水が流れる日もあった。ゴールデンレトリーバーはウォータードッグと呼ばれ、水難救助犬に適した犬種だそうだ。水を嫌がらないどころか、水を見ると飛び込まずにはいられない。雨の日の散歩は特に楽しいらしく、水流に前足を突っ込んで、飛沫をあげて喜んでいる。私もイヌも、雨と汗と湿気でびしょびしょになり、散歩ごとにシャワーを浴びるほどだった。しかし大変なわりに、私の体重は減らず、少し筋肉が増えただけ。残念。

　散歩に出てみて、犬好きな人たちが気軽に声をかけてくるのも、予想外の楽しみとなった。近所にはたくさんの犬が飼われていて、いろんな犬や飼い主と知り合いになった。公園デビューならぬ、お散歩デビューはまずまずであった。

odor in the house was too strong to bear. Though Inu was cute, a smell was a smell.

Inu received three combined vaccine shots including boosters and the first rabies shot at three-months old. We learned he'd need the combined vaccination every year and the rabies shot every three years. On top of that, Inu received the filarial, or canine heartworm, shot at six-months old. We worried that Inu wouldn't like to go to the vet but lucky for us, he seemed to love to visit the good doctor.

So at long last, we went out for a walk together, or at least that was the idea. On the first day, he didn't understand what he was expected to do, so he curled up at our feet, staring up at us like we were crazy. But this didn't last long. By the next day, he followed us with his tail wagging.

We usually walked the steep trail between the Mid-levels and the Peak. That year we had more rain than usual and sometimes water flowed like a river down the steep path. Golden retrievers are known as water dogs and may be suitable for conducting water rescues. They are not scared by water but eager to jump into it, so no wonder Inu enjoyed his walks on rainy days. He loved putting his front legs into the water and making a splash. Inu and I were saturated from the rain, sweat and humidity. After each walk, we needed a shower. Unfortunately, despite the rigor of these walks and the sweating, I didn't lose any weight. My legs were a bit more muscular though.

An unexpected pleasure of walking Inu was that strangers who liked dogs would often talk to us. There were many families with

ここまでの経費：
繰越し金：　　　　　　　　　　　　　　　　¥227,760
三種混合ワクチン（2回）：　　　　　　　　　　¥8,320
狂犬病予防と三種混合、マイクロチップ込み：　¥7,680
フィラリア予防ワクチン　　　　　　　　　　　¥5,760

合計：　　　　　　　　　　　　　　　¥249,520
　　　　　　　　　　　　　　　　　　チーン！

一人で留守番させると不安になって破損行為に出る。

Inu caused destructive behavior from separation anxiety if we left him home alone.

dogs in our neighborhood so we got to know a variety of people and their pets. Just like a baby's debut to the neighborhood playground, our debut to the neighborhood walk went quite smoothly.

Expenditures :	
Balance carried forward :	HK$14,235
Multivalent vaccination (twice) :	HK$520
Rabies, multivalent vaccination and micro-chip :	HK$480
Filarial vaccination	HK$360
Total :	HK$15,595
	Ka-ching!

初めての散歩。うれしそうに座って歩かないイヌ

Inu was happily sitting down at his first walk. It took a while to understand the concept of walking.

43

7．出費はかさむ

　無事に予防接種も終え、マイクロチップを入れてもらって外に
出られるようになったイヌ。寝床がある洗濯部屋にはエアコンが
なかったので、縦置きのものを買った。湿気の強い香港の夏場、
エアコンがないのは大変つらい。だがエアコンをつけても、すぐ
そばにイヌのトイレが設置されていて、湿気と匂いがこもるの
は避けられない状態だった。ところが、散歩に出るようになると、
犬臭さはさっと消え、私たちはほっとした。

　さて、これからが本格的なおトイレのしつけだ。散歩は朝の6
時から夜の9時まで3時間おき、1日6回とした。月齢3ヶ月
を過ぎると、さすがに膀胱が発達してきて、1日10回程度のお
しっこで済むが、代わりに量が増えて、粗相をされると掃除は大
変だ。

　イヌより2歳年上のゴールデンレトリーバー、「マルコス」は
毛並みの美しいオスだ。飼い主は、香港人の家族4人。全員そろっ
て散歩をしている姿はほほえましい。しかもマルコスは、飼い主
に言われるまでもなく、道の脇の草むらに入ってちゃんと用を足
す。7割がたは散歩中に間違いなくできるようになったイヌだが、
マルコスを見ては「将来イヌもああなれるかしら」と羨望と心配
の溜息をつく毎日だ。

　香港特有の事情だが、うちのマンションでは犬を連れている人
は表のエレベーターに乗ってはならない。作業員用の裏エレベー
ターを使うよう指定されている。我が家は45階まであるビルの
10階と、たいして高くないが、引越し業者が入っている日などは、
待てど暮らせどエレベーターがこない。その間に犬がピーとやる。

Chapter 7 - Tired of spending and cleaning

Inu slept in the washing room but there was no air-conditioning in the room. We worried about the summer heat and humidity so we bought a stand-alone cooling machine. Hong Kong summer humidity is a killer. Still, the room smelled like the toilet. But, happily, now that Inu was going out for walks, the rest of the apartment smelled fine.

We began to get serious about Inu's toilet training and set up a detailed schedule. We'd walk him six times a day, every three hours from 6:00 am to 9:00 pm. Now that he was older, he only needed to do his business ten times a day!

Marcus was also a golden retriever with beautiful orange-tan fur. He was two years older than Inu. His owners were a Hong Kong couple with two children. It was such a tranquil sight to see the whole family out walking with Marcus in the neighborhood. Marcus was very considerate. He'd step off the path into the bushes when he needed to go. My husband and I sighed with envy every time we saw that. We hoped Inu could be like Marcus someday.

A unique feature of living in a Hong Kong high-rise was that residents with dogs were not allowed to use the front lift, but had to use the back, or service lift, that was used by delivery men and the cleaning staff. Our apartment was on the 10th floor which was not very high out of forty-five story tower, but whenever someone was moving in or out of the building,

無事外に出ても、住居の敷地部分では用を足してはならない。門を出るまでこらえきれないイヌがまたピーとやると、私たちは守衛に頭を下げて、地面を洗ってもらう。匂いが染みついていると、すべての犬がそこでマーキングを始めるからだ。

　さて、エアコンまで買って、イヌのための臨時出費に悩まされていた私たちだったが、ここでさらに新たなる決断をした。お手伝いさんを頼もうというのである。イヌの散歩を毎日6回こなしてかなり疲れてきたし、休暇には旅行にも行きたい。だが、イヌを買い取ったペットショップのことを考えると、果たしてペットホテルに入れて、イヌが病気をもらわずに出てくるか心配だった。お手伝いさんに住み込んでもらえば、私たちの旅行中、イヌも不安を感じないで済むはずだ。香港の法律では、東南アジアから来るお手伝いさんは住み込みで雇うことが義務付けられており、パートのお手伝いさんという選択肢はなかった。

　それに、この頃からイヌはものすごい勢いで様変わりを始めていた。モヘアのように柔らかい赤ちゃん毛が、尻尾の付け根を起点として抜け始め、代わりに波打つ金色の毛が生えだしたのだ。毎日目に見える成長のスピードで、それは嬉しい驚きだ。だが、家の中に舞うイヌの毛と、まだ粗相が止まらないおしっこのせいで、掃除に明け暮れる私はくたくただった。夫も出社前と帰宅後に、イヌの散歩を欠かさなかった。でも「一日ぐらい開放されたい」というのが私たちの本音だった。体調が悪くても悪天候でも、絶対にトイレしつけのために、外に連れて行かなければならないイヌ。成犬になって落ち着き、1日3回程度の散歩で間に合うようになってからは考えられない大変さが、子犬の世話には要求された。

　住み込みのお手伝いさんという初めての経験に不安もあったが、

we had to wait for a long time before the lift arrived and Inu would, inevitably, pee in the lift lobby. Even when Inu could wait successfully, he had to hold it until we were outside the building gates which was challenging because our housing estate had five towers with a long driveway framed by attractive flower beds. Often, Inu couldn't hold it and we'd apologize to the gate guard and try and wash away the spot because if the smell remained, all of the other dogs would pee in the same spot.

Although we'd already spent a small fortune on items for Inu's care and comfort, we decided to employ a live-in domestic helper. I was exhausted with walking Inu six times a day, plus we wanted to be able to take a holiday sometime. We thought about leaving Inu at the pet shop where we'd found him but were worried he'd get some type of disease or that terrible cough again. We thought a live-in helper would solve all these problems so that we could make travel plans freely knowing Inu would be safe at home. Practically, Hong Kong law requires the employer to sponsor the foreign domestic helper to live-in the employer's house, and there was no option for a part-time helper.

By this time, Inu's appearance had quickly started to change. His soft baby fur started to fall off in clumps and was replaced by wavy golden hair. The speed of growth was so quick and was noticeable every day, which was a happy surprise. But now we had dog hair floating all over the house and he still wet the floor once in a while so cleaning the house had become a big burden. My husband walked Inu each morning before going to the office

私たちは知り合いのつてで、フィリピン人のお手伝いさんを紹介してもらうことにした。

```
ここまでの経費：
繰越し金：                            ¥249,520
エアコン：                             ¥54,400
ヘルパーさん（イヌが１歳までの７ヶ月）： ¥411,040

合計：                               ¥714,960
                                      チーン！
```

毛の長いイヌも暑いだろうが、エアコンは高額。
Air-conditioning is essential for a long-haired dog, but not cheap.

and in the evenings when he returned home but we both needed a day off. Taking care of a puppy was much more work than we'd imagined! Compared to the care you need for an adult dog, a puppy needs way more attention.

Though we were not one hundred percent comfortable with the idea of having a live-in helper, we decided to employ a woman introduced to us by our friends.

Expenditures :	
Balance carried forward :	HK$15,595
Stand alone air-conditioning :	HK$3,400
Payment for helper (until Inu turned one)	HK$25,690

Total : HK$44,685

Ka-ching!

生後4ヶ月ぐらいから生え変わりが始まる。

Around four months old, the puppy coat started to be replaced with adult fur.

8．クレイジーなティーンエージャーとしつけ

　1日6回の散歩と、生え変わり始めた毛の掃除に根をあげて、とうとう住み込みのお手伝いさんを頼もうと私たちに決心させた子犬の「イヌ」。

　実を言うと掃除とトイレのしつけだけが問題ではなく、成長著しい大型犬の子犬が発散する強力なエネルギーに、愛情を感じながらも不安を抱き始めていたのだった。

　人間でもティーンエージャーは扱いづらいものだが、人間年齢で10代に入り、10キロを超えたイヌのパワーはすごいものがある。どういうわけか夜の8時、夫が帰ってくると嬉しさが最高潮に達するのか、ダイニングテーブルの周りをぐるぐる走り出す。走っているイヌにあたられると、こちらに青あざができるほどのすさまじさだ。

　そして、狂ったように（見える）私たちの持っているものめがけて飛びつく。最初はふざけているのだが、イヌは手加減を知らないのでエスカレートする。これも人間のティーンエージャーと同じか。ともかく前足の間に頭を低く下げ、高く上げた尻尾を振りながら「ウウウ」と唸り、ジャンプして私の手の中のものを取ろうとする。故意ではないのだが、手を噛まれることもしばしばだ。

　それだけではない、背の高い鉢植えにアタックして葉をむしり取る。家具に体当たりしてふっとばす。ああ、もうどうしたらいいの。6ヶ月を過ぎたらしつけ教室に通わせるつもりで予約を入れてあるが、そんなことで直るのだろうか。

　外に出るようになった時から、イヌには訓練用チェーン首輪をつけた。散歩の時、あらぬ方向に行こうとするイヌを引っ張ると、

Chapter 8 - Never smack Inu

Inu's arrival helped us make up our mind to employ a live-in helper. We needed help with the walking and cleaning but, to tell you the truth, we started to feel uncomfortable with how strong Inu was becoming. It is difficult to handle teenagers, and dogs are no exception. As Inu was getting into the age equivalent of an early teenager, he gained weight and his power became fearsome.

This was especially true at 8 pm when my husband would return home from the office. Inu suddenly ran like a demon possessed and he'd spin around the dining table. He would jump up in excitement, often causing us to drop whatever we had in our hands. At first, he was just being playful but he didn't know when to stop. He'd growl with his head down in between his legs, his tail wagging high, then he'd jump at whatever I had in my hand, trying to steal it away. Sometimes, he'd bite my fingers although he didn't mean to.

Inu also attacked the tall, potted tree in our living room, damaging all the branches. And he'd smash his increasingly large body into the furniture. But what could we do? We had booked the discipline class for him to begin once he was 6-months old but would it be enough?

For walking Inu, we used a chain training collar around his neck. If Inu tried to pull us in the wrong direction, we pulled the chain. I thought it was a bit cruel but understood it was a necessary

首がグエッと絞まり、言うことをきくという仕組みだ。ちょっと残酷な気もするが、犬と人間が社会生活をともにするのに必要な道具だ。

このチェーンを家でもつけさせてあるのだが、エネルギー爆発時のイヌは、とてもチェーンでは押さえられない。だいたい捕まえられない。大事な書類を引ったくって、めちゃくちゃに噛みちぎった時、私は思わず手をあげた。「この、バカッ」。

振り下ろそうとした私の手を、夫が止めた。「絶対にぶっちゃ駄目だ」という夫の理論は、ぶったって理由がわからないし、全力で遊んでくれると思ってさらにエスカレートするだけ、というものだ。たしかに夫が正しい。でも私は悔しい。こんなに面倒を見ているのに、なぜ言うことをきかないのだ、イヌ。

イヌが1歳を過ぎて少し落ち着いた頃、近所の飼い主仲間の奥さんが、「あそこのゴールデンの子犬、5ヶ月だけれど、しつけをしようとして捕まえそこなって、後ろ足骨折ですって。それも2度目よ」と、耳打ちした。ひどい話だ。絶対に有ってはならないことだ。暴力的にしつけるなんて論外だ。そう強く思う一方、体も力も強くなった犬が、気が違ったように走り回り、家人を恐怖に陥れる様子も目に見えるような気がした。高層マンション住まいでは、ちょっと庭に出すということもできない。大型犬は人間より速く走り、力もある。体重はすぐに30キロに達し、そうなったらもう抱き上げて方向転換をすることもできない。毎日の訓練と信頼だけが一緒に社会生活をする秘訣となるのだ。

成犬になってからは、近所でも評判の性格良しになったイヌだが、あの時私が思いに任せてぶっていたら、すねた犬になっていたのだろうか。

tool to make puppy a good member of society. Inu wore this chain even when he was home but it was useless against his explosive power. In truth, we could not control him, and when he'd steal things from my hands, I couldn't help but raise my hand while screaming "You stupid bastard!" But my husband grabbed my hand. "Never smack the dog!" he said.

He explained that Inu wouldn't understand what smacking meant and might think it was part of playing. Well, he was totally right but I was frustrated. I was doing so much for the dog. Why couldn't he listen to me?

Shortly after Inu turned one, I heard a rumor in the neighborhood from one of the dog owners that a five-month old golden puppy from a nearby apartment had a broken leg twice because the owner tried to punish the naughty pup and failed to capture it. That's a terrible story and I agree that it should never happen. But at the same time, I could imagine how the puppy made the family so exasperated with his strength and boundless energy. Living in a skyscraper apartment, there was no escape to the backyard either. Large dogs are stronger and faster than many humans. But even this can't excuse such brutal behavior by dog owners. Fair and consistent discipline is the key for dogs to live peacefully in human society.

Once matured, Inu was known for his gentle disposition, but every now and again when I lost my temper and smacked him, boy would he become a sulky dog?!

```
ここまでの経費：
繰越し金：          ¥714,960
小型犬用チェーン：    ¥1,120
```
....

合計：   ¥716,080

チーン！

クレイジーに走り回ると顔まで違って見える。
Inu ran like demon possessed at FRAPS
(frenetic random activity period)

Expenditures：
Balance carried forward：　　HK$44,685
Chain collar：　　　　　　　　HK$70

Total：　　　　　　　HK$44,755
Ka-ching!

９．犬の世話あれこれ

　すごい勢いで成長していくイヌという名のゴールデンレトリーバー。成長は嬉しいけれど、大型犬の子犬のパワーは私たちだけでは面倒見切れない。いや、面倒は見られるが、そうすると本来あるべき生活が保てない。

　そこで頼んだ住み込みのお手伝いさんが、８月の半ばにやってきた。名前はローラ。無口でおとなしい人だが、掃除が上手でイヌの扱いもすぐに心得てくれた。たいそう小柄で、イヌに引きずられるのではないかと心配したが、見かけによらず力持ちだ。

　それまでに確立した、イヌの世話をひとつずつ教える。まず、夏場は週に２回のシャワー。ジョンソンのベビーシャンプーを少量使う。耳ダニが治ったので、耳掃除は週に２回で良し。他に歯磨き。これは週に３回。チキン味の歯磨き粉はイヌのお気に入りだ。外から帰ったら必ず足先を拭く。朝昼晩とブラシで毛並みの手入れをする。長毛種なのでブラッシングは一仕事だ。イヌがなつくように、夕飯はローラがあげることになった。他には常に水飲み場の水を入れておくこと、などなどだ。

　イヌの様変わりはものすごい速度で進み、背中に生え出したウェーブが、体中に広がった。特に目覚しい変化は尻尾で、ちょび毛の生えた棒状だったものが、尻尾を巻き込むような強いカールの毛でおおわれ、花のつぼみが毎日膨らむかのように変わっていく。この毛が更に伸びて、成犬になると扇子を広げたような直毛になったのだから、生命が内包する美しさを見せつけられたようで、感動した。

　同時に背も伸び体重も増え、爪も伸びた。これがなかなかと

Chapter 9 - Helper helps but puppy is a puppy

In August, Laura came to work for us as a live-in foreign domestic helper. She was a quite, petite woman. We worried that Inu might drag her down the street but fortunately, she was very strong. We showed her how to take care of Inu step by step. During the summer months, bathe and shampoo him twice a week, using just a small amount of Johnson's baby shampoo; clean his ears twice weekly; brush his teeth at least three times a week for the good dog breath (Inu loved toothpaste with chicken flavor) ; clean his paws each time he returned home from walking; brush him at least three times a day, morning, lunch and evening time (long-haired dogs like Inu need a lot of brushing) ; and, of course, feed him regularly and keep fresh water in his bowl.

Inu's appearance sure was changing. He now had wavy hair along his back and curling around his tail. It was as if buds were blooming every day. As he matured, the wavy curl became like a crescent-shaped fan. His development showed us the mysterious power of life, and we were in awe.

As Inu grew taller and heavier, his nails also grew longer. They were now very sharp so even a light scratch put a welt on our skin. We studied the books to figure out how to cut his nails but we were not confident that we'd be able to hold him in order to do it. We didn't want to hurt his paw by mistake. As a solution, we consulted a dog grooming service. The groomer was a Hong

がった強烈なもので、ちょっとでも触れるとこちらの皮膚にミミズ腫れが走る。本で勉強して自分で爪を切ろうかと思ったが、暴れ回るイヌを押さえつける自信がない。下手をすると肉を切り落としそうだ。

そこでグルーミングを頼むことにした。コニーという香港人の女性で、道具を一式もって自宅に来てやってくれる。15キロの子犬の時から毎月来てもらって、33キロになった今でも同じ料金とは良心的だ。ヘアスタイルだけでなく、皮膚のチェックもしてくれるので、ありがたい。

初めの頃はドライヤーを恐れて、イヌはヒンヒンと泣き声をあげていた。そしてシャワールームに2時間こもってふわふわになって出てきたとたん、お気に入りのカーペットにピーとやるのだ。イヌはきれいになったものの、部屋はおしっこ臭くなるというアクシデントが続いた。が、10ヶ月ぐらいになるとこらえられるようになった。

さて、こんなに急成長しているイヌだが、やはり子犬は子犬で、ピークまで一緒に歩くと、暑さのせいもあり、途中で座り込んで動かなくなる。道路に手足を投げ出して寝そべってしまうので、抱っこして歩かなければならない。これがなかなか重い。

散歩コースには湧き水を石で囲った小さな池がある。イヌは散歩の度にこの水に浸かって体を冷やす。通りすがりの人が「涼めてうらやましい」と声をかけてくれる。水からあがると「シェイク」という号令で、イヌは体をブルリと振り、水滴を撒き散らして毛を乾かす。どうやらローラがしつけたらしい。シャンプーの後、シャワーカーテンを閉めて「シェイク」と号令し、乾かしているようだ。「なかなか利巧な技だ」と、夫と私は感じ入った。

Kong lady named Connie. She came to our house with her tools and equipment. She first cared for Inu when he was just a 15-kilogram puppy and though he was now 33-kilograms, she still charged the same fee. We were happy about that. She also advised us about the condition of his skin.

At first, Inu was scared of the hair dryer and would whine the whole time. When Connie was finished, Inu would race out of the bathroom with perfect, fluffy finish, and pee on the carpet. So when Inu got clean, the room got smelly. This continued until Inu was nearly a year old.

Well, regardless of his quick growth, Inu was still a puppy. When we walked up to the Peak, before we were halfway there, he'd lay in the road, refusing to move anymore. We had to carry him the rest of the way. That wasn't easy.

Along the walkway, there was a small spring encircled by stone blocks. Inu loved to lie in this pond and cool himself down. The passers-by seemed to enjoy this sight. And when he got out, he'd shake his body to dry himself off as soon as we said "shake Inu". Laura taught him that. After giving him the shampoo bath, she'd close the shower curtain and say "shake Inu." My husband and I were impressed with this smart trick.

ここまでの経費：
繰越し金：　　　　　　　　　　　　　　　　　　　　　￥716,080
歯ブラシと歯磨き：　　　　　　　　　　　　　　　　　￥1,840
ベビーシャンプー（半年分）：　　　　　　　　　　　　￥1,920
グルーミング（1回440ドルで5ヶ月から1歳まで）：　￥56,320

合計：　　　　　　　　　　　　　　　　　　　　　　　￥776,160
　　　　　　　　　　　　　　　　　　　　　　　　　　チーン！

棒状にちょび毛が生えただけの幼児期のしっぽ。

Puppy's tail was like a short stick.

ウェーブした毛がしっぽを包み込み全身に広がる。

Then strongly curled hair covered the body and tail.

Expenditures：
Balance carried forward： HK$44,755
Tooth paste and brush： HK$115
Baby shampoo (for half year) HK$120
Grooming service for 8 months HK$3520

Total： HK$48,510
Ka-ching!

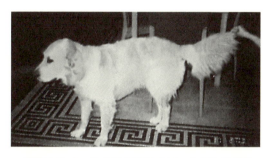

緩やかなカールがある直毛のゴールデンに成長。

Finally, it became a gorgeous fan-shaped tail with wavy fur.

10. 案外かかる食事とおやつ

　イヌの成長に悲鳴を上げ、とうとう頼んだ住み込みのお手伝いさん。彼女が毎日掃除してくれるので、生活は以前の整然さを少し取り戻した。だが一方では、大変ながらも楽しんでいたイヌとの散歩をお手伝いさんに取られ、私は悲しくなった。勝手なものである。

　この頃、イヌが気に入ったおやつに日本製の「ホネッコ」というのがある。甥が日本から遊びに来た時、お土産に持って来てくれたものだ。香港のとあるペットショップで同じ物を見つけ、私たちはイヌを連れて、散歩がてらホネッコを調達しに行くようになった。

　だが、その店のお姉さんはイヌを見るたびに「足が曲がってる」と指摘する。たしかにイヌの足は X 脚だ。獣医さんには、大型犬特有の股関節形成不全だが、痛みはないので、適度な運動を心がけるように言われ、その指示に従って努力していたところだ。

　なのに、足が曲がっていると再三言われ、私はむっとした。「これが犬だから言うのね。私たちの家族なのに。人間の子どもだったら言わないわけ、失礼ね」と怒っていたが、後で友達から、自分の子ども二人を見比べて「どっちが美人か」と評する中国の友人に辟易した話を聞かされた。だからイヌが人間の子どもであっても、言われたに違いない。もっともペットショップの場合は他にも目論見があったようで、「うちなら足の曲がってないもっと良い犬が買えるよ」と言われ、そこにはもう行くのをやめた。犬を商品としか思ってないのだ。ペットショップのくせに。いやペットショップだから、あたりまえなのだろうか。

Chapter 10 - Dog food's expensive!

Now with the helper cleaning our house each day, the house was much more tidy. But I started to miss the days when I was involved with Inu for everything. I felt sad every time Inu and Laura went out for a walk. Human minds are so complex and, sometimes, selfish.

When Inu was 5-months old, my nephew from Japan visited us and brought an edible treat for Inu called Honekko. Inu loved it so much. We found that the local pet shop carried it and we stopped by once in a while to buy some. But the shopkeeper was very annoying. She'd always say to us, "The legs of your dog are not straight!" Okay, that's true. His legs curved in towards his middle. The vet said Inu had mild hip-dysplasia which was apparently common for big dogs. In Inu's case, he didn't have any pain from it. The vet recommended we walk him a lot and we followed his advice. I complained about the shopkeeper to my friend, "The lady at the shop said she can convert Inu because he's a dog, not a human child. But Inu is like our child so how rude she is!"

My friend was very annoyed too because one of her Chinese friends had the audacity to compare her two daughters to judge who was prettier. She did this right in front of the girls. So, she said even if Inu was a human child, some people would stop to point out his defects. That may be true but I think the pet shop had something else in mind because the shopkeeper always said

イヌは我が家に来て以来、乾燥したドッグフードであるドライフードを食べていた。獣医さんに勧められた子犬用を1日3回食す。6ヶ月頃から食事は日に2回となり、おやつとして、スーパーでも手に入るデンタボーンという、硬くて歯に良い骨のようなものを日に2、3本食べるようになった。他には犬用ビスケットを10日で一袋ほどあける。ジャーキー類やウェットフードは、アレルギーが強いゴールデンには向かないと獣医さんに言われたが、実際、ジャーキーをあげるとイヌは下痢をしてしまう。なかなか繊細だ。

　去勢をした後は、成犬用のドッグフードに移行したが、どのブランドでも下痢が止まらず、一時は「万一肝臓が悪いといけないので」と検査を受けた。が、肝臓は正常で、人間のアトピーのような過敏症だった。IVDというブランドの「鹿肉とジャガイモ」という、特製ドッグフードでイヌの下痢はぴたりと止まった。これにウォルサムのノーファットのドッグフードを混ぜる。鹿肉9キロとノーファット4キロを約2ヶ月で消費する。ここに到達するまでに買い換えたブランドは約5種類。一度食べると下痢をするので、そのたびに余りをご近所の犬に差し上げた。まったくお金と手がかかるったら。獣医さん推薦のブランドはスーパーで売ってるものより2倍近く高いが、新たに試して下痢をされると困るので、高いものを買い続けている。

　アトピーのお子さんを持つお母さんに聞いたところ、家畜の肉だとアレルギー症状が出やすく、鹿など野生動物の肉なら大丈夫なことが多いのだそうだ。イヌでもアトピーは大変なのだから、アトピーのお子さんを持つ親御さんの苦労は相当だろう。

　もっとも人間の子供は下痢をしたら自分でトイレに行ってくれるが、イヌの場合は静かにしてしまう。泡立つ黄金の液体が白い

that she'd sell us a better dog with straight legs for a good price.

We stopped going to that shop. They treated dogs like products even though they were running a shop for the care of animals. Or maybe they thought dogs were like products precisely because they were running a pet shop.

We had always fed Inu dry dog food. He received specialized puppy food three times a day. At 6-months old, we only had to feed him twice a day, plus two or three Dentabones a day for treats (and good teeth) . Inu also consumed a bag of dog biscuits every ten days. Our vet said that wet dog food and the beef jerky treats were not recommended for golden retrievers because they often had some allergies to foods. This was true. Anytime we gave Inu jerky, his poo got runny. He had a sensitive stomach.

After Inu was neutered, we switched to the adult specialty dog food, but this gave Inu diarrhea. We tried many different brands recommended by our vet. They were expensive, but time and again, Inu had the soft pooh, so we gave the premium food to our neighbor's dog. This happened for so many weeks that the vet did a test on Inu's liver to see why he had this reaction. Fortunately, his liver was normal and the loose stools were just a sensitive reaction.

When we switched his food to the IVD brand of "deer and potato" – which could be purchased at the vet's office as specialty food and was much more expensive than the usual ones sold at supermarket-his diarrhea stopped instantly. We added the Waltham brand, no-fat dogfood. He consumed nine kilograms

カーペットに沁みこんでいくのを、鼻を抑えながら拭き取るのは大変な作業なのだ。

```
ここまでの経費：
繰越し金：                                    ¥776,160
デンタボーン（一袋約22ドルで一歳まで）：          ¥63,360
クッキー（一袋約55ドルで一歳まで）：             ¥15,840
子犬用ドッグフード（260ドルで3～6ヶ月）：         ¥12,480
合わなかったドッグフード（買い換え五種類）：       ¥14,240
IVDドッグフード（390ドルで6～12ヶ月）：          ¥18,720
ウォルサムドッグフード（250ドルで6～12ヶ月）：    ¥12,000
肝臓検査代：                                    ¥6,400
```

合計：   ¥919,200

チーン！

足が曲がってたって、かわいいイヌ

His curved legs didn't affect his cuteness.

of "deer and potato" and four kilograms of the no-fat food in two months. A friend of mine who had a child suffering from an atopic disorder explained that often livestock meat causes an allergic reaction whereas wild meat such as deer causes fewer allergies.

My friend, she must have been very busy caring for her child because it was a lot of work getting the problem under control even with Inu. On the other hand, at least a human child can trot to the toilet. Inu, he scurried to the corner of the room, leaving a puddle of foamy golden slime soaking into the white carpet. I had to pinch my nose to clean it up; it smelled so bad.

Expenditures :

Balance carried forward :	HK$48,510
Dentabone（$22/bag untill one year old）:	HK$3,960
Cookie（$55/bag untill one year old)	HK$990
Puppy food $26/bag for 3 to 6 months)	HK$780
Disagreeable food（5 different brands)	HK$890
IVD dog food（$390 for 6 to 12 months)	HK$1,170
Waltham no-fat dog food（$250 for 6 to 12 moths)	HK$750
Liver examination	HK$400

Total :   HK$57,450

Ka-ching!

11. かわいそうだけど、去勢

　イヌは案外繊細なお腹を持っているのが、しばしばの下痢症状で確認された。が、最初に長期の下痢をしたのは、精神的負担が原因だったように思う。

　9月に入り、イヌが5ヶ月半になる頃、私たちはのびのびにしてきた里帰りで、1週間家を空けた。お手伝いさんにイヌもなつき、このあと、去勢、しつけ教室と押しているイヌのスケジュールを考えると、この時をはずして出かけられないと思ったからだ。

　明日は香港に戻るという日、お手伝いさんから「イヌのウンチが柔らかい」と電話が入った。翌日、1週間ぶりに会ったイヌは、興奮さめやらず、3時間ほど家の中を走り回った。この時ばかりは私たちも一緒に走りたいほどだった。何しろ旅行の間、イヌのことが気にかかり、持参した写真を会う人ごとに見せて、ヒンシュクを買っていたのだから。

　帰宅した翌日をピークに、イヌの下痢は快方へ向かった。初めて留守番をさせられて、イヌにしたら置き去りにされた不安があったのだろう。「1週間で帰るから」とどんなに説明してもわかってもらえないのが、つらいところだ。

　イヌは5ヶ月の頃、遊びにきた7歳の甥っ子に恋をした。立ち上がったイヌとほぼ同じ背丈、イヌより2キロほど少ない体重。その甥っ子にのしかかっては腰を振るという行為を、イヌは繰り返した。その後も毎日成長して、ますます抑えがきかなくなってきている。去勢はかわいそうとも思ったが、香港のマンションで一緒に暮らすには必要なことだ。イヌにとってもホルモンの調整が効いて癌にかかる率が減り、そのほうが幸せだと本に書いてあ

Chapter 11 - Neutering Inu

In September, when Inu was five and a half months old, we took a trip to our home country. We'd postponed the trip many times until we felt confident leaving Inu at home. Finally, we decided Inu was used to our house and the new helper.

On the last day of our trip, we received a telephone call from Laura saying Inu's pooh was runny. The next day when we returned home, Inu was so excited to see us that he raced around in the house for three hours. We were happy to see him too, though we didn't run. During our travels, we couldn't stop thinking about Inu and showed his photos to everyone we met, which was a bit embarrassing now that I think back on it.

Inu's runny pooh got better the next day. He must have been anxious without us. It meant the stomach trouble was caused from the psychological reasons, not the allergy. I wished he could understand our verbal explanation that we'd be home in just a week.

Though still a puppy, Inu fell in love with my 7-year old nephew who visited us. The two were the same height when Inu reared up, though Inu weighed a few kilograms more. Inu kept mounting on my nephew's leg which was a little awkward so we thought the time had come to neuter him. We felt sorry for him, but we knew it was necessary for us to live together in the skyscraper apartment in Hong Kong. Plus, the book said that neutering may help to stabilize hormones and reduce the risk

る。そう納得し、旅行から帰った翌週、予約を入れると獣医に連れて行った。

　朝入院すると、夕方には退院で、手早いものだ。ところで、イヌはもうひとつ手術をした。2週間前から突然上唇に出現したできものを切り取って、悪性ではないか検査してもらったのだ。1回の麻酔で2個の手術が終わればイヌも楽、という理論だ。

　退院したイヌは、エリザベスカラーと呼ばれるラッパ状のものを首からつけていた。口の傷を守るためと、去勢した傷を舐めないようにとの用心だ。

　にもかかわらず、唇の傷は2日ほどで、糸が抜けてぱっくり割れてしまった。手術の痛みを感じているとは思えないパワーで走り回り、顔をぶつけるからだ。獣医さんもお手上げで「いまさら縫っても」ということになり、イヌの口にはそれ以来ひだのような傷がある。面白いのは手術のために顔の毛を剃ったら、毛は白いのに、出てきた皮膚は鼻と同じく真っ黒だったことだ。剃った毛が生えそろうと、唇の傷は大して見えなくなった。

　イヌはエリザベスカラーを嫌がって、前足を引っ掛けて取ろうと努力していたが、半日ぐらいで諦めた。食べ物や土がつきやすいカラーの手入れは大変だったし、何より抱き寄せた時の冷たいプラスチックの感触は嫌なもので、1週間後にこれが取れた時、イヌも私もほっとした。できものも悪性ではないことがわかり、二重にほっとした。

　ただ、手術後もイヌの傍若無人な元気さは、なんら変わることがない上、去勢の直後は局部が異常に腫れて大きくなっているので、本当に取るべきところを取ったんだろうか、と傷跡を見ながら私たちは首をかしげていた。

of cancer so we convinced ourselves it was the right thing to do and booked the appointment for the next week.

We took Inu to the vet in the morning and picked him up that evening. It was not a big deal. We also asked the vet to examine the blotch on his upper lip that had appeared two weeks earlier. He had two operations with only one anesthesia, so we thought that was a good idea. He came out with the plastic, Elizabethan collar shaped like a cone surrounding his head to protect the cut on his lip and prevent him from licking the incision for the neutering.

Despite this, in two days the wound on his lip opened because Inu kept banging his face against the furniture so the protective bandaging came off and the stitches came out. The vet decided it would be a waste to stitch it again so Inu was left with a dented scar on his lip. It was pretty visible with his skin having been shaven for the operation but once his fur grew back, we could hardly see the scar. Also, we thought it was funny that once he was shaved, the skin under the fur was as black as his nose.

Inu really hated the Elizabethan collar and tried to pull it off with his feet. It was pretty sad to watch but, thankfully, he finally got used to it. It was hard to keep the collar clean because dog food debris and dirt from the street easily got stuck in it. Most of all, I hated the cold plastic feel when I held Inu. When the vet took the collar off, we were all relieved. The doctor concluded the blotch was not malignant, which made us even happier.

Inu's hyperactivity didn't change after the neutering and within a few days, his private parts swelled, leaving us to wonder if the

```
ここまでの経費：
繰越し金：           ¥919,200
去勢手術代：          ¥14,400
おできの手術代：        ¥8,000
皮膚癌検査代：        ¥12,800
·····················································
合計：             ¥954,400
                    チーン！
```

恋するイヌは抱きつきたいお年ごろ

Inu couldn't stop jumping on my nephew because Inu fell in love.

operation had been done properly.

Expenditures:	
Balance carried forward:	HK$57,450
Neuter operation:	HK$900
Blotch operation:	HK$500
Examination for skin cancer	HK$800
Total:	HK$59,650
	Ka-ching!

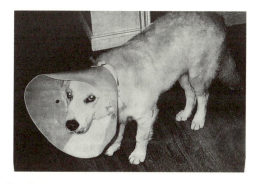

英国貴婦人の襟飾りを連想させるエリザベスカラー

Inu didn't look noble with Elizabethan collar

12. しつけられるのは人間

　去勢手術の腫れが引いた頃、イヌは4週間のオベディエンス・コース、つまり、しつけ教室を受けることになった。やんちゃなどというかわいいものではないイヌのパワーに引きずられ、散歩のたびに自分が転ばないか心配している私たちは、このコースに大きな期待を寄せていた。

　訓練士の名前はジャック。若い頃、香港の警察犬の訓練で、自ら噛まれる役を買って出て、ノウハウを学んだというつわものだ。その後カナダで正式に学校にも通い、今では多くの顧客を持ち、獣医さんからも推薦されるほど定評がある。贅肉のない精悍な体をしているが、イヌの訓練を始めてみて、その理由がわかった。訓練は引き綱を引く人間にとってもなかなかハードなものなのだ。

　ジャックは毎日夕方来て、1時間のトレーニングを地下駐車場で行う。この時、私、お手伝いさんのローラ、会社から急いで帰ってきた夫の3人も出席する。まずイヌに「歩け」（ヒール）、「お座り」（シット）、「ふせ」（ダウン）などの基本動作を覚えこませる。そして、私たちに引き綱の持ち方からイヌとの歩き方、号令のかけ方を教えてくれる。

　ジャックが指示すると、イヌは2日めには要領を得て、3日めには基本の指示を飲み込むという優秀さを発揮した。

　なんだカンタンじゃない、と思ったが、それは大きな誤りだった。ジャックの言うことを聞くイヌは、私たちの指示には従わないのだ。引き綱をもって上手に歩くのは案外難しく、方向転換などで息が合わないと、イヌを引きずったり引きずられたりする。

Chapter 12 - Disciplining people rather than the dog

We put Inu in a four-week obedience course once the swelling had subsided. We really hoped the course would work because Inu was getting stronger everyday and was becoming difficult to walk on a leash because he pulled so hard. He'd become so powerful, we worried we might get hurt from walking him.

The trainer was named Jack. He was a war-horse who volunteered to train police dogs in order to learn about dog training. Later, he studied in Canada to obtain a certification and he established a good reputation in his profession. He had many customers in the neighborhood and our vet recommended him for Inu.

Jack had a trim, masculine body and we discovered why after we started Inu's training. Jack came to us early evening for a one-hour training session in the parking area in the basement of our apartment building. Our helper and I joined the training, as did my husband when he returned from the office. First, Jack trained Inu to understand basic orders such as "heal", "sit" and "down." Then he trained us on how to hold and handle the lead and how to give orders to Inu.

Inu was a quick learner and he mastered all the basic instructions by the third day. I thought this had all worked out really well but I was mistaken. Although Inu listened to Jack so obediently, he didn't follow any of our instructions. It was more difficult than it looked to walk with the dog while handling the

私、夫、ローラと三人三様の歩き方に左右されるイヌもかわいそうだが、こちらも汗だくだ。一番優秀なのがローラで、彼女には遊んでもらえないと分かっているイヌは、おとなしく一緒に歩く。当然一番駄目なのが、イヌを甘やかしてきた夫だ。歯をむき出して吠え、「一緒に遊ぼう」の態度を取り、引き綱を引っ張り、夫に飛びつく。まったく手におえない。「ボスじゃなくて友達だと思ってますね」とジャックに診断されてしまった。

　「歩け」に該当するヒールは「バイ・マイ・ヒール」の略で、「私の踵につけ」という意味なのだそうだ。つまり、イヌが人間より前を歩いていてはいけない。指示を聞かない時は「駄目」（ノー）といいながら、一度たるませた引き綱を強く引くと、首のチェーンが食い込んで、「これはやってはいけないんだな」とイヌが理解する。そのはずだが、この引き綱をたるませるのがけっこう難しい。十分たるませて体に衝撃を与えるように引っ張れば、イヌはきちんと従う。だが、たるみが少なく、引き綱の引っ張りっこになると、自分の方が強いことを示そうとして、イヌは全力を出す。

　ジャックの前では優等生面をして、私たちの言うことは聞かないイヌを見ると、面白くない。でも、こちらの引き綱の扱いが下手なのだから、イヌをうらむのは筋違いというものだ。訓練が終わったあとも、「言うこときかないと、ジャックを呼ぶからね」とイヌに言い聞かせている。もちろん言葉を理解しないイヌは、涼しい顔である。

lead in the proper manner. If you and your dog were not very coordinated, it could be difficult. It was a pity for Inu who had to get used to three different walkers – all of whom were beginners -- but we tried our best. Laura was the best with him because Inu knew she would not be playful with him so he didn't fool around. And, naturally, my husband was the worst because he had been spoiling Inu by giving in to his demands. Inu barked and growled, baring his teeth to my husband, which meant "play with me." Inu played tug-of-war by pulling the lead, and then jumping on him, out of control.

"That's because he thinks you are not the master but a friend," Jack explained to us.

I learned that "heal" was the abbreviation of "by my heal" and that it meant that dogs shouldn't walk ahead of people. I also learned that when a dog didn't follow the instruction, I was meant to say "no" while simultaneously sagging the lead then pulling it hard so that the lead gave a certain impact on the dog's neck. By doing this, the dog understood this was not the thing to do. Well, that's the theory anyway. But, in real world practice, it was difficult to sag the lead. If we were quick enough to get the impact on his neck, Inu followed the instruction; but if we failed to sag the lead enough before pulling it, Inu started to play tug-of-war and was out to prove that he was stronger.

It wasn't very interesting for us to see Inu be a model student in front of Jack when we knew he wouldn't listen to us at all. But it was not fair to blame Inu for this because the reason he didn't follow us was because we were not good handlers. After

ここまでの経費：	
繰越し金：	¥954,400
オベディエンス・コース4週間：	¥128,000
引き綱（大型犬用皮製）：	¥4,000
大型犬用チェーン：	¥1,280
合計：	¥1,087,680

チーン！

大きくなっていろいろできるのはいいけれど・・・
Was it good or bad he had grown to reach anything?

we finished the training course, every time Inu got too wild, we scolded him by saying, "If you don't listen to us, mommy will call Jack." But Inu just ignored us.

Expenditures :
Balance carried forward :   HK$59,650
Four-week obedience course :   HK$8,000
Lead (big size leather) :   HK$250
Chain collar for big dog: HK$80
....

Total :   HK$67,980

Ka-ching!

13. 犬の行けるところ、行けないところ

　しつけ訓練の成果も出始めた7ヶ月の頃、イヌの体重は25キロを上回った。こうなるとかなりの距離を歩けるようになる。訓練のおかげで、私たちも完全ではないながら、イヌを座らせたり待たせたりの指示ができるようになり、休日の散歩が楽しくなってきた。

　日曜日は香港トレイルなどを歩いてトレッキングする。トレイルは自然に満ちていて、起伏に富んだ剥き出しの岩や土の細道が多い。こういうところでは犬の能力が人間を遥かにしのぐ。まごまごしている私たちの横を、風のようにイヌは走りぬける。私たちと前後しながら一緒に歩くことはできるが、細い道でイヌと並んで歩くのは不可能だ。

　だが、香港の規則では体重20キロ以上の犬は引き綱をつけなければいけない。小さいうちは人間が犬を引っ張って歩くことも可能だが、25キロを過ぎると完全にこっちが引っ張られる。岩の上で引っ張られて転んで頭を打ったら、と考えると恐ろしい。引き綱を手に巻きつけたままイヌに引っ張られると、手の甲に青あざができるほどだが規則に従わないわけにもいかない。ジレンマだ。

　それに香港には「大きい犬」というだけで怖がる人が、当時はまだ多かったようだ。大型犬を連れていると「キャアア」といって避けていくが、その声でイヌが驚き興奮する。犬は匂いを嗅いで、相手を認識しなければ気がすまない。そばによって鼻を近づけると、実際は触れてもいないのに怖がる人がたくさんいる。ましてや引き綱を離していたら睨まれる。こちらはぺこぺこ謝りな

Chapter 13 - Some places dogs can't go

At nearly seven-months now, Inu began showing some progress from the obedience course. He now weighed over 25kilos and was ready for long walks. Thanks to Jack, we were getting better at instructing Inu and our walks had improved.

On Sundays, we went trekking on the many trails on Hong Kong Island. They were abundant of nature, steep inclines and declines, and narrow dirt paths with rocks and other outcroppings. Inu was faster and more graceful than us which presented a problem.

Hong Kong regulations require that all dogs over 20kilograms must be leashed when out in public. While Inu was small, we could control him by the leash, but once he was more than 25kilos, he was the one pulling us on the narrow path. This was dangerous because if he pulled too hard, I thought I might trip and hit my head on the rocks. This really scared me. And even without tripping, I had to loop the lead around my wrist several times, which bruised my hand. But still, we had to follow the regulation and that was a big dilemma.

Besides, back then, many local people were scared of "big dogs."As Inu approached, many locals would run away screaming, but that made Inu excited and he would try and chase after them. Also, dogs like to smell people to identify who they are, but many of the locals didn't like being sniffed at by Inu.

The worst thing was that in that first year, Inu couldn't stop

がらイヌを必死で捕まえる。

しかも困ったことに1歳ぐらいまで、イヌは道で出会った人や犬に、頻繁に飛びついた。これは自分がやられても嫌だろうと思う。トレッキングの場所では汚れても良い服を着ている人が多いから、たいてい許してくれるが、本当にバツの悪い思いだ。

一方、香港のすばらしいところは犬をタクシーに乗せられることだ。小さい頃からしょっちゅう車に乗せて、慣らすようにした。子犬の間は酔ってしまうので、空腹時に乗せた。それでも山道の多い香港なので、最初の2回は吐いた。静かだな、と思ったら密かに吐いて、目に涙を溜めていた。5ヶ月になると酔うこともなくなり、タクシーでトレッキングの中間点まで行き、そこから歩いて帰れるようになった。日曜になるとイヌはタクシー待ちの列に座り込む。遠くへ連れて行けという意思表示なのだ。

また、離島へ行くフェリーにも犬を乗せることができる。ただし、口輪（マズル）をする事になっているが、一緒に行けるところがあるのは嬉しい。早速ラマ島へ行き、初めての海を体験させた。水の好きなイヌは海に入ってお風呂に浸かるように座っていた。どうやら自分が泳げるのを知らないようだ。ピチャピチャ水を舐めて、しょっぱいのに驚いたらしい。ただし普段と違う地域へ連れて行った時は、地元の犬や野犬に気を配らないと、吠えられたり後をつけられたりすることもある。

どうしてラマ島まで行くのかというと、香港のビーチは犬禁止なのだ。たいていのビーチには犬にバツのついたマークがかざしてある。もっとも監視員のいないオフシーズンや早朝は大丈夫だが、犬嫌いの人がいると嫌な顔をされる。悪いのはこっちなので何も言えない。公園などでも犬は禁止だ。いたるところに犬禁止マークがあって、香港には、案外犬と走り回れるような場所はな

jumping up on people. It was pretty embarrassing. I was constantly apologizing for his behavior. The good news is that on the trekking, most people wore casual clothes so they accepted our apology even though Inu left spots on their clothing. It was pretty embarrassing.

A wonderful thing about having a dog in Hong Kong is that you can put dogs on the taxi seat next to you by paying the luggage price. We gave Inu a lot of rides when he was little because we wanted him to get used to the car ride. At first, he was very queasy, and because the Hong Kong roads are so windy and mountainous, he threw up on us the first two times. After that, we made sure his stomach was empty before he had a ride. We wondered why Inu was so quiet in the car, then noticed the taxi floor was all wet and he looked so sad. By five-months old, he got used to the ride so we drove by taxi to the middle point of the trekking route and walked home from there. On Sundays, Inu went to the taxi lane of our apartment building and sat in the queue, meaning he wanted to go for a ride.

Ferryboats between the many islands in Hong Kong also allowed dogs provided they wore a muzzle. It was nice that you could make a day trip somewhere with your dog. First, we went to Lamma Island and Inu experienced the beach for the first time in his life. Inu loved his baths at home and seemed to think the sea was a big bathtub. He sat in the shallow water as if he didn't know he could swim. He licked at the water and was surprised that it had the salty taste. When you take your dog to somewhere new, you'd better be aware of local dogs and stray dogs that

いのである。

ここまでの経費:
繰越し金:　　　¥1,087,680
口輪:　　　　　　　¥1,120
....
合計:　　　　¥1,088,800
　　　　　　　　チーン!

「愛犬と渚を走る私」という夢を壊す犬禁止マーク
"No dogs allowed" signs crushed my dream of running along the beach with my dog.

chase after your dog, who's a stranger in their neighborhood.

The reason we travelled to Lamma Island was because most beaches in Hong Kong Island did not allow dogs. Nearly every beach had a sign posted with a picture of a dog with a line slashed through it. Off-season and early morning before the lifeguards arrived, you could get away with it but if a non-dog lover was around, they would frown at you or sometimes say something unpleasant. You couldn't say anything in return because you were the one violating the regulation. Most urban parks were off limits to dogs, too. With so many "no-dogs-allowed" signs, there weren't many places to take your dog and fewer places still where dogs could run around freely.

Expenditures :
Balance carried forward : HK$67,980
Muzzle : HK$70

Total : HK$68,050

Ka-ching!

14. 親友フェイジャとマズル事件

　イヌを連れて歩くようになってみると、意外と犬禁止の公園や
ビーチが多いのに改めて驚かされた。もっとも普段はお決まりの
散歩コースを歩いて、ご近所の犬やその飼い主と仲良くなってい
る。

　飼い主の中には自分の犬がよその犬とくっつくのを嫌がる人も
いる。ノミ治療をしていなかったり、皮膚病にかかっている場合
が多い。イヌが子犬の頃は「ノミ治療をしてますか？」とはっき
り尋ねる人もいた。また「去勢してますか？」とも聞かれる。イ
ヌは４ヶ月になった頃から、毎月１回、ノミおよびダニ退治の
投薬をしている。６ヶ月で去勢もした。

　散歩で親しくなった中でもイヌの一番の親友はジャーマンシェ
パードのメス、フェイジャだ。イヌより２ヶ月年上で、ころあ
いも良い。人の通らない裏山の道で２匹を放してやると、いつ
までも追いかけっこをしている。

　大型犬がじゃれあうのは、驚くような光景だ。てっきり喧嘩し
ているのではないかと思うほど、全力でぶつかり合う。相手を倒
して、本気ではないが耳を噛んだりする。負けたと思うほうは即
座におなかを見せて、服従のサインをする。

　うちのイヌは相手が大型犬でなくても、すぐにお腹を見せて服
従するので、近所の飼い主に評判が良い。ゴールデンレトリーバー
の特徴なのだろうが、穏やかで我慢強い。といっても本気で遊ん
でいる時はすごいパワーだ。くんずほぐれつで、ぶつかったらこっ
ちが怪我をする。フェイジャはイヌより一回り大きいだけでなく、
走りが違う。シェパードの敏捷な走りはカッコよくて狼を思わせ

Chapter 14 - Feijoa and the muzzle

With so many places off limits, our daily walks were limited to a couple of regular courses so we met the dogs and their masters in our neighborhood. Some owners didn't like their dog to get to close to other dogs because they were worried about fleas or skin rashes. When Inu was a puppy, some owners asked me, "Is your dog treated for fleas?" or "Is your dog neutered?" Inu started flea and tick treatment at four-months old and was neutered at six-months old but some owners still didn't want their dog to get too close to Inu.

We did make some friends though. Inu's best friend was Feijoa, a female German shepherd. She was two months older, and they got along very well. We'd take them on a narrow path behind the mountain road and they'd chase one another all around.

It was amazing to see the big puppies getting so playful with each other. They would wrestle one another to the ground hard, chewing against one another's ears until one off them capitulated by rolling over on the back as a sign of capitulation.

Inu seemed to always be the one lying on his back but that gave him a good reputation in the eyes of the other dog owners. Golden retrievers were generally quite gentle and patient, but when they played with other dogs, they generated such a force of power that we dared not get too close. They tussled and wrestled and bumped against each other. Feijoa was bigger than Inu, and also faster and more agile. When she ran, it reminded

る。一方ゴールデンは華やかな尻尾をちぎれるほど振り、左右に揺れて走る。もって生まれた能力の差で、決してフェイジャに追いつくことはできない。と分かっているが、子供が徒競走で1等を取れない親の心境とはこんなものなのだろうか、とちょっと残念になる。

フェイジャの飼い主はアメリカ人のご夫妻で、フェイジャに普段から厳しい訓練を施している。遠く離れていても「伏せ」の号令を発して服従させるのには感嘆させられる。

ある日、マンションに1通の回覧板が回ってきた。犬の飼い主全員に当てたもので「今後はすべての犬に口輪（マズル）をはめるように」というのだ。犬の飼い主達は皆、裏口を使い、迷惑をかけないよう気を使っているので、更なるお達しにかなり頭にきた。その話題で飼い主達が盛り上がっていると、フェイジャのマミーがやってきて、「あれはフェイジャのせいなの」という。散歩をしていたら、突然、5、6歳の男の子が車から降りてきた。前を通ったその子の匂いをフェイジャが嗅ごうとしたところ、男の子の母親が「あの犬が私の息子を噛もうとした」と、騒ぎ出したのだという。

犬の習性を知らないから、狼に似たフェイジャを見て、やみくもに怖がるのだろう。たしかに危険なこともあるが、世界中の犬に口輪をはめるわけにはいかない。犬にどう対処するか教えるのが先決だろう。しつけられた犬でなく、野犬だったら騒ぎ立てるのは本当に危険だ。

マンションの管理局に「子どもたちが犬について学びたいなら、うちのイヌは噛まないので、お貸ししますよ」と話したが、口輪騒動はいつのまにか立ち消えになった。

us of a wolf. On the other hand, when Inu ran, his body sashayed left and right as though pulled side to side by his full tail. Inu could never catch up to Feijoa. She had the natural talent for running whereas Inu had a nice big tail. We knew that, but felt a little disappointed sometimes, like how a parent might feel if their child never won at the track and field races.

Feijoa's owners were an American couple who regularly gave Feijoa discipline training. We were impressed every time Feijoa followed the instruction of "down," hollered at her from a distance.

One day, we received a notice in our apartment building addressed to all dog owners. It said "All the dogs are requested to wear muzzles from now on." All of the dog owners in our building took great care to not disturb residents by following the regulations such as using the cargo lift from the back door, always leashing the dogs, and so on, so we were upset by this notice.

One day, several of us were chatting at the backstairs and Feijoa's mom came to us and admitted it was because of Feijoa. When they were walking one day, a boy of five or six years old exited from a car and passed in front of Feijoa so she tried to sniff the boy, and that's when his mother hollered, "That dog tried to bite my boy!"

The mother must have not known about the behavior of dogs and that's why she was so scared by Feijoa. I couldn't say that all dogs were safe and wouldn't bite a young boy crossing their path, but can you muzzle all the dogs in the world? Better if the

ここまでの経費:
繰越し金:　　　　　¥1,088,800
ノミ薬（4〜12ヶ月）:　¥5,760
....
合計:　　　　　　¥1,094,560
　　　　　　　　チーン！

気の合う犬たちの飼い主同士も気が合うことが多い
When dogs are getting along, so are the masters.

child's mother had helped her son learn about dogs and their nature.

We talked to the Management Office, suggesting that they might want to organize a class for children to learn about dogs, even offering that Inu participate as the test subject dog because he never bit people, but there was no response.

Expenditures :
Balance carried forward :   HK$68,050
Flea medicine（4 to 12 months old）:   HK$360

Total :   HK$68,410

Ka-ching!

15. 香港を震えさせた犬の毒殺事件

　世の中には犬嫌いの人だっているのだから、迷惑をかけないように暮らそうと心がけているつもりだ。が、なかなか難しい時もある。フェイジャのように良くしつけられた犬でさえ怖がられ、マンション中に「マズル（口輪）着用」のお達しが出るのだから。

　もっともこんな事で自尊心を傷つけられたように感じること自体、私たちの頭が「犬の飼い主モード」に設定されてしまった証拠だ。「うちの犬はこんなに良い子」と感じ「こんなに努力してちゃんとしつけているのに」と思うわけである。周りが見えなくなって、親ばかならぬ「犬ばか」になると危険だ。気を引き締めなければならない。

　それに、我が家のイヌはマズルを嫌がるけれど、これをはめるのはある意味で正しい。噛み癖や吠え癖がある場合もそうだが、犬は路上のものを何でも食べるので、それを止めさせる意味でも有効だ。

　このことを痛感させる出来事が香港に起こった。犬を狙った毒殺事件である。犯人は除草剤などを鶏肉に仕込み、道に撒く。誤って食べた犬は突然倒れて苦しみだす。すぐに嘔吐させて40分以内に適切な処置を施せば、助かることも多いと聞くが、救急車には乗せてもらえない犬を、どうやって短時間で獣医に運べば良いのか。意識を失った大型犬は、とても一人で動かせるものではない。

　この毒殺事件は1995年に最初の被害報告が出され、それ以来130頭以上が被害に遭ったと聞く。2002年の初めにはボーエンロードに「下毒人」とかかれた犯人のメッセージが赤ペンキで書かれ、世間を騒がせた。現在でも犯人逮捕に至っていない。イヌ

Chapter 15 - Poison shakes the community

We understood that not everyone in the world loves dogs so we worked hard not to bother other people. But it wasn't always easy. We had to admit that our brains were set to "dog owner" mode – much like a proud parent – so we felt humiliated by the mandate to put a muzzle on our dog. Dog-owner brain mode felt like "our doggy is the best" and "we were successfully doing all our best to keep him disciplined," and as a result we were blinded by these thoughts.

Although Inu hated wearing the muzzle, it was wise to protect others, and in addition to preventing barking and biting, it protected Inu from eating things on the street. This became essential when a terrible incident happened in Hong Kong. An evil person was killing dogs by leaving poisoned food on the street. The perpetrator was adding poison such as chemical herbicides to meat. If a dog ate it, he would collapse on the street. If he vomited immediately and was taken to the veterinarian within 40 minutes, he might survive. But how could we carry a big dog to the veterinarian? Ambulances refuse to carry dogs, so if this happened to us, an unconscious Inu would be too heavy for us.

The series of dog poisoning incidents was first reported in 1995 and since then over 130 dogs had been victimized. At the beginning of 2002, Hong Kong was further shaken to see the message written by "dog poisoner" in red ink on the rock wall

が9ヶ月のころ開かれたSPCA（愛護動物協会）の「飼い主と被害者の会」に私たちは参加し、被害者や刑事さんの話を聞いた。もどかしいのは犬が殺されても「所有物の破損」という罪状にしかならないことだ。被害者やSPCAの必死の呼びかけにもかかわらず、直接人間に害が及ばない事件は後回しだ。しかも犬を殺す目的で路上に肉を撒いたかどうか、判定が難しく、犯人のめぼしがついても逮捕には結びつかないという。

　だが犬を毒殺された被害者たちの声は痛切で、人間の家族を奪われたのと同等の嘆きだ。SPCAや警察の捜査が成果をあげないことに業を煮やし、懸賞金5万ドルを個人的に設置して犯人逮捕を促す人もいた。夫はこの捜査のために、SPCAに5000ドルを寄付した。けちな性分なのか、私は、それは多すぎると思った。

　が、SPCAでは無料で嘔吐剤を飼い主たちに配り、対処法についてネットで情報を流すなど、懸命の努力をしている。応援するのは当然だ。その上不思議なことに、犬を1匹飼うと、どんな犬を見ても愛しいという気持ちが湧くようになってしまう。SPCAで里親を待っている犬や近所のきれいな飼い犬は言うに及ばず、路上生活の犬でさえ、尻尾を振って無垢なまなざしを投げかけてくると、私はもうメロメロだ。もし自分のイヌが毒殺の被害にあったらと考えるだけで、涙がにじみ出る。5000ドルで助かるなら御の字なのかも知れない。

　この会合に出て以来、散歩の時に携帯電話と嘔吐剤を持参することにした。食い意地の張っている我が家のイヌが、路上の鶏肉を見逃すはずはない。しかも子犬は何でもかんでも試してみないと気がすまないのだ。イヌの散歩は突然緊張を増した。

of Bowen Road, the street on which many of the poisonings had occurred. Yet, still, the perpetrator was not captured.

When Inu was nine-months old, we joined the "dog owners and victims" meeting held by SPCA and heard speeches from the victims' families and police detectives. We were disheartened to learn that even if your dog was killed, the most the criminal could be charged with was for "damaging property." Given this, police couldn't make these crimes a priority over cases affecting harm to people, regardless of the well-articulated appeals from the SPCA and the victimized families. In addition, it was very difficult to prove if the poisoned meats were specifically intended to kill dogs. So although the police had a suspect in mind, it was not easy to arrest the person.

The cries from the victimized families were fierce, and sorrowful as losing a child. Some became enraged by the inaction of the police in capturing the culprit and organized a private ransom of HK$50,000 to promote the capture of the criminal. My husband donated HK$5,000 right after the meeting. I thought that was a bit too much but I agree I'm a bit stingy by nature.

But I had to admit that the SPCA did important work by distributing nausea pills (free of charge) to dog owners and in spreading detailed warnings on the internet. We knew we needed to support these activities. And besides, once you became a dog owner, you started loving all dogs, not just neighbors' pretty dogs and dogs waiting for adoption at the SPCA, but even the dogs living on the streets who could melt your heart with a wag of their tail. The idea that Inu could be poisoned made me

ここまでの経費：
繰越し金： ￥1,094,560
SPCAへの寄付金： ￥80,000
....

合計： ￥1,174,560
チーン！

大きくなっても抱っこは大好き、でももう限界。
Inu loved to be carried, but too big for me to carry anymore.

so upset. If $5,000 HKD could help save a dog, the price was cheap.

Ater the meeting, we carried our mobile phones and the nausea pills on our walks just in case. Inu had a big appetite and would never miss out on any meat left on the road. Puppies would eat anything! So we took great care in walking Inu and tried our best to keep him safe.

Expenditures：	
Balance carried forward：	HK$68,410
Donation to SPCA：	HK$5,000
Total：	HK$73,410
	Ka-ching!

「悪事の報いを受けよ」といった意味らしいけれど・・・
The poisoner's red-ink warning of bad karma to the people.

16. トイレは完璧になったけど

　幸運にも我が家の散歩コースに毒入り団子が撒かれることは無かった。「犬毒殺事件」が世間を騒がせた間、緊張こそしたものの、相変わらずの毎日だった。

　と思ったが、気がつくと、イヌは家の中で粗相することがなくなっていた。イッチョマエに片足をあげて、ピーとやるようにもなった。ウンチの時はスクワットの体勢に入ったイヌのお尻の下に、さっと古新聞を差し込む。手もイヌも汚さず、さっと丸めて捨てるだけ。香港の素晴らしい点で、道路のあちこちに犬の糞専用のオレンジ色のごみ箱が設置されている。1日に30回もおしっことウンチを拭いていたことを考えると、感動的な成長だ。

　10ヶ月を過ぎるころ、もう1回ジャックのしつけ教室をたのんだ。人間なら15歳ぐらい。もうなんにでも飛びつく子供っぽさは無くなったが、体重は30キロ近くに増え、自分の行きたい方へ引っ張る力はますます強くなった。

　久しぶりに会ったジャックは「よく運動しているようで、足がまっすぐになった」とイヌの成長ぶりをほめてくれた。ちょっと鼻が高かったが、これは毎朝出社前にピークまでイヌを連れて往復する、夫の努力によるところが大きい。

　だが、一方では夫のおかげで再度ジャックを呼ぶことになったのだ。最初の授業でも夫はイヌのしつけが一番下手だったが、ジャックがいなくなるとイヌ同様、自分の好きなように散歩をするようになった。いくらローラと私が、引き綱を短くもって「ヒール」と号令をかけながら散歩をしても、夫と自由気ままにやっているイヌは言うことをきかない。散歩の途中で突然「遊びモー

Chapter 16 - Although his toileting became perfect...

Lucky for us, we didn't encounter any poison meat and we were able to continue with our same routine. That routine had gotten much better because Inu had grown up and could control his toileting. He now raised his leg to pee and waited to go poo until we were on the street, and we could then squat down and place a newspaper under him to catch his droppings and place them in the trash. Many trash boxes specialized for dog poop were installed alongside the streets in Hong Kong, which were fantastic. Thinking back on the period when we had to clean up to 30 times a day, this was a big improvement.

After Inu was 10-months old (about 15 in human years), we asked Jack to come back for a second round of discipline classes. It wasn't that he was jumping on everything but that he was getting stronger everyday and he'd pull the leash in the direction he wanted to go (which wasn't always where we wanted to go). Jack praised us when he saw Inu's legs had straightened which reflected that he was getting enough exercise. We prided ourselves on being good parents but it was mainly due to my husband who walked Inu every morning to the top of the Peak before going to his office.

On the other hand, it was also due to my husband that we had to call Jack again. He didn't do any of the things Jack told him to do in terms of holding the leash and getting Inu to heal. In fact, he let Inu walk freely as though Inu were in charge. That meant

ド」に入ると、私の握っている引き綱に飛びつき、端を咥えて「引っ張りっこ」をしてくれと唸る。そして獲物を咥えたライオンみたいに首を振り回し、引き綱を持つ私たちをも振り回す。その力の強いこと。暴れ回るイヌを抑えられない私に、周囲の視線が冷たい。猛獣使いを避けるみたいに通って行く。「あなたは力があるからイヌに負けないでしょうけど、私とローラにとっては本当に危険なの」と、さんざん文句を言った。だが二人、もとい、夫とイヌはうまくやっているらしくてらちがあかない。そこでジャックの登場となった。

今回はさらに「取って来い」（フェッチ）や「待て」（ステイ）も訓練メニューに入れた。ジャックの指導のもと、イヌは前に教えられたことをちゃんと覚えていることを証明した。つまり私たちの指示に従わないだけだ。「真剣な顔で真剣な声音を出して」というジャックに従い、おなかのそこから声を出す。イヌが指示に従わない時は、イヌの行動の先回りをして、引き綱をたるませ思い切り引っ張る。「待て」の状態のイヌが、待ちきれずに走り出すと、急いで走っていって引き綱を掴み、定位置までイヌを引きずり戻す。イヌの定位置を中心に、行ったり来たり走り回るのは人間だ。「こんな競技が運動会であったなあ」などと考える私の額を、汗が流れる。この年になってこんなに走らされるなんて。練習には原っぱを使ったので、大いに蚊に刺された。

大きな瞳に微笑んでいるような口元。誰が見ても「カワイイ」イヌなのに、コントロールを失うと凶暴にすら見える。「2歳になると落ち着くから」と他の飼い主たちがアドバイスをくれた。今になってみるとその言葉は正しいのが分かる。人間にも落ち着く年齢というものがあるではないか。だが当時の私にそんなことは考えられず、自分の手には負えないものを背負い込んでしまっ

that whenever Laura and I tried controlling the leash by holding it short and pulling it while commanding "heel," Inu just ignored us or, worse, he'd suddenly change into "play" mode. When that happened, he'd bite on the leash and start a game of tug-of-war, pushing me back and forth and side to side. I could barely hang onto the leash. People moved aside as though avoiding the lion tamer who'd lost control of his beast. I complained to my husband a lot, insisting that he learn to control Inu so it wouldn't be so dangerous for the helper and me. But he listened to me about as well as Inu did, so I had to ask Jack to come.

This time the training included "fetch" and "stay," and Inu proved that he had remembered everything he learned from Jack in the first session. So that meant Inu simply didn't listen to our orders. Jack urged us to make "the serious face" and to give "the serious voice"; and if he didn't follow our order, we had to sag and pull the leash hard to give him warning. If Inu couldn't hold in the "stay" command long enough, we had to run ahead of him and drag him back to the "stay" position. So when Inu was supposed to "stay," that meant I'd be running back and forth with sweat running down my face. I'm not that young anymore so this was quite a physical exertion for me. Plus, I got bug bites all over because I worked on this on a small, outdoor field.

Nobody could deny that Inu was cute with his big, round eyes and a grin on his mouth, but he wasn't tame. Other dog owners advised us that he'd calm down around the age of two. They were right, of course, but I didn't know that then and feared our Inu would never settle down.

101

たのではないか、と恐怖を覚えたのだった。

```
ここまでの経費：
繰越し金：                    ¥1,174,560
ジャックのしつけ教室（二週間）¥64,000
...........................................................
合計：                        ￥1,238,560
                                   チーン！
```

トイレをマスターしたら、早速夫の隣で寝るようになったイヌ。

Once he mastered toileting,

Inu decided to sleep next to my husband on the bed.

Expenditures：
Balance carried forward： HK$73,410
Discipline course（2 weeks）： HK$4,000
....
Total： HK$77,410
Ka-ching!

「取って来い」は覚えたけれど、獲物は絶対放さない。
He mastered "fetch" but never "release" game.

17. 暴れん坊の親の気分？

　再度ジャックのしつけ訓練を受け、力あり余るイヌを何とかコントロールしようと努力していた私たちだが、8ヶ月ぐらいから12ヶ月ぐらいのイヌは、お手上げ状態だった。人間なら12歳〜18歳ぐらい。別に、グレたり薬物に手を出すというわけではないが、仲間の犬を見つけると、引き綱を振りきって走り出し、仲間と一緒に人間には追いつけない速度で走り回る。一種の暴走族だ。

　イヌの親友はジャーマンシェパードの雌、フェイジャだが、もう1頭の仲良しが、ロットワイラーの雄、カイザーだ。この2頭はゴールデンよりさらに大型。一緒に遊んでいる光景は、くんずほぐれつの大乱闘をしていたかと思うと、狼の俊敏さで走り出すという、ローラースケートをつけたプロレスみたいなすごさだ。2、3ヶ月若いということもあり、2頭に比べれば小柄な種であるイヌは、一緒に走っても置いてけぼりのように私には見える。

　だが、本人は仲間と一緒が嬉しいらしい。この2頭が遊んでいる真ん中に、身を投げ出すように飛び込んで行くのだ。当然「降参」のポーズをした挙句、体中を「噛み噛み」されて帰ってくる。やめれば良いのにと思うが、それでもイヌにとっては楽しいらしい。

　だがある日、11ヶ月になろうとするイヌは、夫に連れられ、後ろ足に3箇所の大きな歯型をつけて帰ってきた。2箇所は針を刺したように深く出血している。が、一番問題なのは膝に該当する関節の上についた、盛り上がった傷だ。

　獣医に連れて行くと、麻酔で眠らせて、骨接ぎのように関節を

Chapter 17 - Parents of a rough child?!

Inu became challenging when he reached the teenage years (8-12 months old in a dog's life). He wasn't delinquent or into drugs but he was hyperactive. Whenever he saw his gang for friends, he'd careen towards them, jerking us about as we struggled to hang onto the leash. Inu's best buddies, Feijoa and Kaiser, a male rottweiler, were like a Japanese bike gang when they stampeded together. They'd spring and scuffle one another to the ground like professional wrestlers with roller skates. he was a couple of months younger and the smaller of the three but he didn't seem to care. He'd dive right into the middle of the other two dogs, ultimately "giving up" to them. I didn't like it but Inu sure thought it was fun.

One day when he was almost 11-months old, he returned home from a walk with my husband dragging his rear leg. When I looked more closely, I could see three teeth marks, two of which were quite deep and bleeding. He also had a wound on his knee joint that was swelling.

I took him to the vet and he gave him anesthesia to sleep and straightened the leg joint. Inu didn't wake up for long time after the procedure so the vet gave him the medication to wake up. When he came to, his rear legs were too weak to stand. It was really heartbreaking.

We wanted Inu to learn from this experience, but the next time he saw Feijoa and Kaiser, he leapt into the middle of them again.

引き伸ばしてくれた。麻酔がなかなか覚めないので、覚ます注射をしたが、意識が戻っても足腰が立たず、半日以上もヒンヒンとか細い声を出していた。歩けない犬というのはなさけないものだ。本人はかなり不安だったようだ。

　これで少しは懲りたか、と思ったが、痛みを感じない体質なのではと疑うほど、フェイジャやカイザーを見ると飛び込んでいって遊びたがる。傷が癒えるまでは、こっちの手がちぎれるほど引っ張られようと、絶対に他の犬と遊ばせないことにした。

　それにしても、自分の犬が傷つけられたとあっては、心中穏やかではない。でも「どの犬が噛んだかなんて、くんずほぐれつの中、とても見分けがつかない」と夫は言うし、そもそもたのまれて一緒に遊んだのではなくて、２頭が仲良くしているところにイヌが飛び込んでいったのだ。非はイヌにある。そういえば小学校時代、体が大きくて運動神経の発達した男の子たちが遊んでいる中へ、あとさき省みずに小さな体で飛び込んでいった挙句、怪我をして、周りに迷惑をかけている男の子がいたな、と私は思い出した。安藤君といったかな。怪我をさせたとされる男の子の親は、謝りながらも釈然としなかっただろうな、と思う一方、でも、やっぱり一言フェイジャとカイザーのマミーに言っておきたい、という欲求を抱えて、私はいらいらした。

　１ヶ月ほどで傷は癒えたが、またしてもイヌはフェイジャとカイザーの中に飛び込み、同じような場所に切り傷をつけて帰ってきた。前ほどひどい傷ではなかったし、あっちこっちに擦り傷や切り傷をつけるイヌに慣れっこになってきた私は、市販の塗り薬をつけてやった。「獣医さんは高いんだから。毎回は行けないの」と言いながら。

106

We had to pull him back from the playing because he hadn't healed yet. Inu was not happy with us, pulling on the leash so hard in protest that the leash nearly cut into my hand.

At any rate, we were not happy that our dog had been hurt by the other dogs even though we couldn't be sure which dog bit Inu because they all rolled over one another so much. And besides, they did not ask Inu to play. It was Inu who jumped into their game so we knew it was Inu's fault. It reminded me of times playing in primary school when the smaller boy got hurt when playing with the more athletic, bigger boys. You want to say something to the parents because a child was injured, just like I wanted have a word with Feijoa's and Kaiser's mommies about Inu's injury. But what could I really say?

In a month, his wound was healed and, once again, he jumped into the middle of the dogs, ending up with cuts around his legs. This wasn't as bad as the last time and, by this point, we were so used to Inu getting injured that we just put the antibacterial ointment on his wounds to save having to pay the vet every time.

ここまでの経費：
繰越し金： ¥1,238,560
獣医： ¥12,800
....
合計： ¥1,251,360
　　　　　　チーン！

怪我した足に包帯。いい加減に懲りてほしい。
Dragging his damaged leg, but Inu never stopped jumping into play with his friends.

Expenditures：
Balance carried forward： HK$77,410
The vet for treating injury： HK$800

Total： HK$78,210
Ka-ching!

18. 留守にしてごめんね

　自分より大きな犬の中に飛び込んでいって、じゃれあっては怪我をする我が家のイヌ。「ジステンバーなら死ぬかもしれない」といわれた2ヶ月のころのひ弱な面影は、どこへ行ったのやら。急成長の時期が過ぎた後、腰の位置が徐々に高くなり、逞しい成長を続けた。

　だが、気持ちはまだ子供。誰かが一緒にいないと吠えまくり、こちらはトイレにも行けないという状況こそ収まったが、「家族は一緒」という主張は相変わらずだ。

　私たち夫婦は2ヶ月に1度ぐらい、所用や休暇で出かける。小さな香港なので、出かけると言えば必ず海外旅行となる。その間、イヌはローラとお留守番だ。すると必ずなるのが下痢症状。アレルギー症のせいもあるが、イヌに合ったドッグフードを探し当てたため、普段は大丈夫だ。ところが私たちが旅行で留守にすると、とたんにゆるくなる。これは精神的なものらしい。

　私たちの帰宅で、イヌの下痢は治るが、そのあと徐々に「自分一人を置いてきぼりにした」ということを思い出すらしい。帰ってきた直後はとても喜ぶのだが、三日目ぐらい、私たちが毎日家にいることを確認したあたりから、ヘンな行動を始める。夜中の2時に吠え出したり、散歩から帰ったとたんに裏階段に逃げ出したりする。最上階は45階だから、探すのは大変だ。だが、イヌにとっては楽しいゲーム。置いてけぼりにされたことへの腹いせのようでもあるし、私たちがついてくることを確認しているようでもある。単に寂しかったショックで、いつものルーティーンから外れてしまうだけかもしれないが。

Chapter 18 - Family should always be together

Inu continually got injured from playing with bigger dogs but that's what he wanted to do. We didn't worry so much because he was getting bigger and sturdier all the time. It was hard to believe this was the same dog we met at 2-months old who was so weak that the vet wasn't sure he'd survive.

But he was still a puppy. While he no longer barked every time we prepared to leave, his strong desire for the family being together was unchanged. About bimonthly, my husband and I went out of town on business or holiday and since Hong Kong is such a small country, that meant travelling overseas. We'd leave Inu in our helper's care but he always suffered from loose poo in our absence. We bought him special food that was easier on his stomach but he had a psychological problem when we were away.

As soon as we returned, Inu no longer had the soft poo, and he was so happy to be reunited with us. After a few days, though, he usually started acting out from resenting that he'd been left alone. He'd bark at 2:00 am or race up the back staircase past our apartment. There were 45 floors in our building so this was a problem. But these were his playful games, his subtle revenge for us having gone away without him, or maybe for him it was a confirmation of our bonding because we had to catch him. Or maybe there was no reason at all for his odd behavior.

One Sunday a few days after returning from Tokyo, Inu ran away

散歩中のイヌが突然逃げ去ってしまったのは、私たちが東京旅行から帰って3日目の日曜日だった。裏の道から階段を上ったところに小さな貯水所があり、原っぱになっている。出口は1ヶ所しかないので、いつものようにそこでイヌを放して走らせていた。と、突然、葉が落ちた冬の木々の間を、イヌは悠然と走り去ったのだ。人間には降りられない崖だが、成長したイヌにはお茶の子さいさいだ。もっともこの頃には、イヌは私から離れても、すぐに戻ってくるようしつけられていた。だから私は待っていた。

　ところが、10分待っても20分待ってもイヌは帰らない。これは尋常でない。でもどうしたら良いのか分からない。声を限りに「イヌ、イヌ、イーヌ」と叫びながら、とりあえずマンションへ帰った。守衛さんにイヌを見たかと聞いても、知らないという。

　教会から帰ってきた夫とともに、再度探しに出ると、さっきとは違う守衛さんが、「33階に預けたのはイヌかも」という。33階にはチャタムという13歳になるゴールデンレトリーバーの老犬がいて、ジョーというフィリピン人ヘルパーさんが、日曜日も欠かさず面倒を見ている。案の定、33階で、クッキーをもらってチャタムとくつろぐイヌを発見した。ジョーにお礼を言って200ドルの心づけを払い、イヌを引き取った。

　安心すると同時に憤慨もした。食い意地の張っているイヌは、猫好きな人たちが野良猫に配る食べ物を狙っていたのだ。原っぱの崖下に、その匂いをかぎつけたに違いない。目撃した人によれば、猫を蹴散らしてキャットフードを食べていたそうだ。

　でも、普段はしない行動に出た理由は、もちろんそれだけじゃないのだ。ハイハイ、5日間も留守にして、私たちが悪うございました。

from us while walking. There was a small grassy area above some rocky steps on the backside of our building where the water tank was stored. We thought it was a pretty contained area so we'd let Inu off his leash to run around freely. But that day, he suddenly sprinted through the trees and up a steep, rocky mountainside. I was surprised but tried to remain calm, calling for him to return to me. I waited but Inu didn't come back. After 10, then 20 minutes had passed and he still hadn't returned, I started to panic. I frantically called his name and dashed back to our apartment in the hopes that somehow he'd come back down the hillside without me seeing him. But he wasn't there, and the guard at our building said he hadn't seen him.

When my husband returned from church, we set out together to search for him but another guard advised us that he'd taken a golden retriever that matched Inu's description to the 33rd floor. Chatham, a 13-year old golden retriever, lived there with a Philippine helper named Joe who was always taking good care of him, even on Sundays. We raced up to the 33rd floor and found Inu eating cookies with his new buddy Chatham. We thanked Joe very much and insisted she take $200 for having cared for Inu.

We were so relieved but, at the same time, angry at Inu for having put us through that stress. We learned later that Inu had chased some wild cats up the hill and eaten some food that had been left for them although we suspected that the real reason was because we'd left Inu at home to go to Tokyo.

ここまでの経費：	
繰越し金：	¥1,251,360
迷子のお礼：	¥3,200
合計：	¥1,254,560

チーン！

木々の間を軽々と走り抜けるイヌ。
Running through the hillside bushes
and trees easily where people can't go.

Expenditures：
Balance carried forward： HK$78,210
Gratuity： HK$200
....

Total： HK$78,410
Ka-ching!

おいてきぼり。家族に思いをはせている、ようにも見える。
Longing for the family to come home.

19. 遊びとお気に入りおもちゃ

　私たちが旅行に出ると、機嫌を損ねるイヌだが、私たちだって気を使っているのだ。その証拠にちゃんとお土産を買ってくる。だいたい、夫婦二人で旅行しても、会話はイヌのことしかないから、自然とペットショップに足が向かってしまうのだ。

　さて、私たちがイヌを飼うことになってからでも、香港のペット事情は格段に良くなった。人々がペットという考えに慣れ、動物愛護運動などの高まりで、一部の人だけでなくみんなが、犬や猫を受け入れるようになったと感じている。ペット用品を扱う店も増えた。それでも、当時は品揃えが豊富とはまだ言えず、特に大型犬のおもちゃはあまり選択肢がないと感じた。

　やはりペットショップが充実しているのはアメリカだ。香港シティーホール３階の中華料理店ほどもある巨大な売り場に、これでもか、というバラエティに並んでいる。品質も良い。イヌの場合、一番困るのが、どんなおもちゃでも１週間で噛み切ってしまうことだ。ほぼ２週間おきに犬用ぬいぐるみを買い与えていたが、アメリカで買った大型犬用の羊のぬいぐるみは、犬がくわえて振り回しても５ヶ月もった。

　イヌの好きな遊びは「引っ張りっこ」と「取ってこい」だ。床に置かれたおもちゃ箱から、好きなものをとってきて、私たちに差し出す。それを握ると、イヌが引っ張る。すぐに放すと、「本気で引っ張れ」というように唸る。本気で引っ張って、おもちゃを奪い取ると、イヌの目が期待に輝く。そこで狭い家の中、できる限り遠くに投げてやると、喜んで走っていって、取って帰ってくる。そして同じことを繰り返す。こうやって引っ張りっこをす

116

Chapter 19 - Japanese pet shops are interesting

To allay our guilty feelings when we were out of town, we always brought Inu a gift. And even though we were away, my husband and I seemed to talk about Inu a lot so we found ourselves seeking out the pet shops.

In a short time, we observed that the environment for pets in Hong Kong was steadily improving. More people were accustomed to the idea of having a pet as a member of the family and so the number and size of pet shops had increased dramatically. But still, the line of products was rather limited, especially in terms of the selection for big dogs.

In contrast, pet shops in the United States were incredible. Many were as large as the Chinese restaurant on the third floor of City Hall, and they carried such an abundant variety of products. And the quality was high, too. In Hong Kong, we bought soft toys for Inu every other week because Inu chewed them to pieces. But the sheepskin toy we bought in the U.S. lasted more than five months even though Inu chewed it everyday and swung it all around.

Inu loved to play "tug-of-war" and "fetch."He'd select a toy from his box on the floor and push it to us. We'd grab one end and Inu would pull from his end. If we gave up too easily, he'd growl in a dissatisfied way. When we swiped it away, his eyes lit up. We'd throw it away as far as we could in the room and Inu would dash to fetch it. We'd do the same thing again and again. Inu could do

るのも、おもちゃが長持ちしない理由である。

　香港はアメリカやオーストラリアの輸入品が多いが、日本は独自のペットグッズ市場があるようで、日本のペットショップを覗くのは楽しい。とてもおしゃれなものがおいてある。インテリアになる犬用品とか、自然素材を使ったおもちゃとか。でも私が回ったお店では大型犬用のグッズがあまりなかった。お土産に買ったヘチマのひも付きおもちゃは、イヌが１日で噛み切ってしまった。

　イヌが一歳になる３月、妹が日本からプレゼントを送ってきた。黄色い星型のクッションだが、なんと防弾チョッキの素材で作ってあると言う。お店の人に、「ゴールデンちゃんなら噛み切ってしまわれることもあるので、こちらがよろしいのでは」と勧められたのだそうだ。このクッションはイヌがどんなに噛みつき、引っ張っても壊れない。店員のアドバイスはばっちりだったわけだ。

　お礼を言う私に、妹は「何でペットショップの人って、犬に敬語を使うのかな」と面白いことを言った。たしかに、子ども用品売り場の人が、買い物客の子どもに対応するのと同じように、丁寧に話しているのだろうが、つい敬語になってしまうのがおかしい。

　「この犬、足が曲がってる。うちならもっと良い犬が買えるよ」と香港のペットショップで言われた私たちだ。日本なら、さしずめ「こちらのゴールデンちゃんは、お足がまっすぐとは言えませんが、私どもではまっすぐの犬も用意してございます」などとなるのかな、と勝手に翻訳してみた。そういう言い方なら、あんまり気分を害さないだろうか。でも、日本のペットショップでは「犬を買い換えろ」なんてことを、そもそも言わないに違いない。ばかな翻訳を考えるのは、やめにした。

118

this for a very long time.

Hong Kong pet shops had many products imported from the U.S. and Australia but we discovered that pet shops in Japan had some unique products such as toys made out of organic materials. But the organic toy made out of loofah only lasted a day.

In March, when Inu was going to be one year old, my sister sent a birthday gift from Tokyo. It was a star-shaped, bright yellow cushion that seemed to be made out of a bulletproof material. The shopkeeper had apparently recommended it for golden retrievers because it was known to last. Inu played with that star cushion everyday, chewing and pulling and tossing it about, but it never tore. That shopkeeper knew what he was talking about.

When I called my sister to thank her, she asked me an odd question. "Why do shopkeepers use honorific expression to mention dogs?" Well, maybe it's the same tact displayed in children's clothing stores where the salesman addresses the children in a polite and respectful manner. I thought of the saleswoman in Hong Kong who tried to convince me to buy a dog with straighter legs. Maybe in the Japanese pet shop, he'd have the legs of a nobleman. Who knows, but I don't think the Japanese pet shop would tell me to get a better dog.

ここまでの経費：	
繰越し金：	¥1,254,560
おもちゃ（1歳まで）：	¥8,960
お土産（3回の旅行）：	¥5,600
合計：	¥1,269,120

チーン！

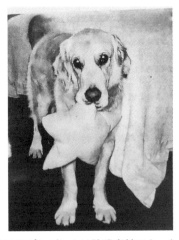

誕生日のプレゼントは防弾素材のクッション

A birthday gift cushion made out of bulletproof material.

Expenditures :
Balance carried forward : HK$78,410
Toys till one-year old : HK$560
Gifts from the travelling : HK$350

Total : HK$79,320

Ka-ching!

20.「だから飼えない」VS「それでも欲しい」

　現在イヌは 2 歳、体重は 35 キロになった。この本を読んでみると、「そういえば子犬の時は嵐のように大変だったなあ」と思うほど、落ち着いた犬に成長している。イヌは長時間寝る。たいてい敷物と化して、寝転がっている。もうそんなに手がかからない。が、どんなに大変でも、小さくてホワホワしていて、抱き上げることのできた、子犬の時代が懐しい。

　1 歳になるまでをつづってみたが、経費の高さに自分でも驚いている。79,320 ドル（1,269,120 円）とは。本当にこんなにかかったのだろうか。「だから犬は飼えないの」という言い分を擁護するために、あまりにも高額に書きすぎていないだろうか。

　そこでかからないかもしれない経費を考えてみた。お手伝いさんの給与 25,690 ドル（411,040 円）をイヌの経費に入れるのはおかしい、と指摘されるかもしれない。私がしっかり掃除をすれば済むことだ。しかも現在の落ち着いたイヌを見ると、たいして部屋を汚したりしない。だが、お手伝いさんがいなければ、旅行の度にペットホテルを頼むことになる。一泊 270 ドル（4,320 円）ぐらいかかるらしい。年に 30 日旅行すると 8100 ドル（129,600 円）。そこでもらった病気の治療代が 2000 ドル（32,000 円）、というのは憶測だ。もちろん病気をもらわないかもしれない。

　他に大きく削れるのは SPCA への寄付金 5000 ドル（80,000 円）。気持ちとしては 500 ドル（8,000 円）ぐらいにしたい。他に食費、家族と同じ素材を自分で料理したほうが、安いのか高いのか。

Chapter 20 - "That's why you can't!" vs. "Still I want one!"

Now that Inu was 2-years old, he was a calm and nice dog. He slept many hours, lying on the floor like a furry rug. He didn't need nearly as much care from us anymore. It was nice but at the same time, I missed the days when he was a puppy always vying for our attention and we could carry him in our arms.

I looked back on all the receipts for the vet, the pet shop and all the other expenses and was shocked to see how expensive it was. HK$79,320 in just one year alone! Did it really cost that much? Maybe I'd miscalculated to support my position.

So I reviewed the costs for items that perhaps not everyone would incur. Some may say that our helper's salary shouldn't be in there. If I cleaned the house myself, there'd be no need to have a helper. And, besides, once Inu was grown up, he hardly made a mess around the house. While these are fair points, if we didn't have a helper, every time we travelled overseas, we would require a pet hotel to keep Inu for us. That's probably HK$270 a night for about 30 days a year, or HK$8,100. But then you have to factor in the additional medical bills for things like kennel cough that he'd pick up from all those other dogs.

You may also question the HK$5,000 donation to SPCA or suggest that we give Inu leftovers from our meals rather than the expensive specialty food suited to Inu's sensitive stomach. These are all fair points.

Some of these costs are one-off costs in the first year such as the

初年度のみかかる経費の代表は、イヌの値段そのもの。これももらえればただ。子犬は予防接種も多い。2年目以降はこれほどかからない。イヌの医療費は保険が利かないが、成犬になると丈夫になるので、医療費もかなり減る。

　さて、お金に関してはこのくらいにして、私が犬を飼ってよかったと思っているかどうか。これは難しい問題である。なぜなら答えが簡単すぎる。もし答えるなら「イヌのいない生活なんて考えられない」ということになり、「だから犬は飼えないの」という言い分を擁護できなくなってしまう。これではこのエッセイの主旨に反する。

　でもしょうがないではないか。体当たりされたり、引きずられたり、高額なお金がかかったりしても、「イヌがいなければ体験しなかったはずの時間」を経験してしまった。私を見ると本当に嬉しそうに甘えた表情で寄って来るイヌ。いったいこの地球上の他に誰が、私にこんな表情を見せてくれるというのか。

　でも、多くは語らない。「ほら、やっぱり、飼うのを勧めてるんでしょ」と言われると、エッセイが失敗ということになってしまうからだ。

　そこで、私がどう思うかは置いておいて、中学生と高校生の母親であり、イヌと同じゴールデンレトリーバーのセイラちゃんのマミーの言葉を紹介したい。「家族4人、大喧嘩をしてむっつりしている時でも、テーブルの下で、交互にセイラをなでてるんです」と言う。そう、犬には求心力といったものがある。あの「みんな一緒にいなければならない」というパワフルな主張は、崩れかける人間関係を安定させる力を持っている気がする。

　ああ、やっぱり駄目だ。「飼ったほうが良いと思ってるんでしょ」と言われそうだ。だからもう何も言わない。でも、ここま

price for Inu. It's possible that you might get a dog free from a dog rescue facility. And certainly it's true that puppies require many vaccinations but once a dog is fully grown, there may be fewer medical expenses.

Aside from the financial cost, if you asked me whether having a dog was good for me or not, the answer is so clear. We can't think about life without Inu! This is troubling given my thesis so maybe this whole essay is a bit of a bummer. But what can I do? Although it was expensive for us and we got jumped on and dragged about, we had so many wonderful experiences with Inu and did so many things we wouldn't have otherwise done if Inu were not in our lives. Inu runs to me in joy every time I walk through the door. Who else on the earth would show me affection like that?

But I should be careful not to talk too much about how happy I was with Inu because I don't want you to feel like I'm saying you must have a dog. So I'm going to leave you with something a friend of mine said. She has two teenagers and a golden retriever named Sara, and when I asked her if she thought having a dog was a good idea, she said, "Even when the kids are fighting with each other, or with me and my husband, everyone is patting Sara under the table."

So, you see, dogs have a power that's like a centripetal force to unify the family. Their insistence on being together prevents the fraying of relationships that so often happens with people. So maybe my essay was not successful since, in the end, I support being a dog owner. But even so, and even if you raise a beautiful

で来るには、おしっこやウンチを1日30回拭いたり、雨の日も風の日も散歩に出て、手にあざができるまで引っ張られたり、転んで怪我をしたりするのを乗り越えなければならない。その上に79,320ドル（1,269,000円）、ということだけもう一度言っておきたい。

　それでも犬が欲しい、という人は、どうぞご自由に苦労を背負い込んでください。

こんなに小さかったイヌ
Inu was this small when he came to us.

dog like Inu, don't forget what I told you about cleaning the poo and pee up to 30 times a day, walking on stormy days, getting pulled and jumped and dragged along and spending gobs of money.

After all this, will you say to your child or spouse, "That's why you can't have a dog in Hong Kong!" or, despite the troubles, "I still want one!" It's your decision but just don't say I didn't warn you.

こんなに大きくなったイヌ
Inu had grown up to be a beautiful dog.

あと書き

　イヌは2014年2月25日、13歳を目前に、穏やかに家族にみとられて、眠るように亡くなりました。犬の血液の病気であるティック・フィーバーに感染し、高額な人間の対HIV薬を投与しましたが、完全に元に戻ることはありませんでした。9ヶ月延命しましたが、衰弱し、度重なる脳溢血に苦しんだ時期もあります。ですが、最後までお手洗いの迷惑をかけず、毅然として私たちに愛情をふりまきながら天国に召されました。ありがとう、イヌ。お疲れさまでした。

Postface

Inu passed away on 25th February 2014 without waiting his 13th birthday. We were at his bedside when he fell in sleep peacefully at the night, just he didn't wake up anymore. He suffered from tick fever, which is a bacterial infection transmitted by ticks. The disease is fatal, especially for older dogs. We treated him with an expensive anti-HIV medicine that extended his life for nine months, but, he never recovered from the weakness and was repeatedly suffered from strokes. Despite that, he was always lovely and proud, not to trouble us with the toilet mistakes even after being suffered from the tick fever. Thank you Inu, for being so good to us! We will always miss you!

プロフィール

リンゼイ美恵子

1961年、東京生まれ。青山学院大学経営学部卒業後、外資系銀行勤務を経てフリーランス通訳翻訳。1991年、イギリスのシェフィールド大学にてＭＢＡ（経営学修士）、2011年、香港大学にてＭＦＡ（文学修士）を取得。1996年、中央公論社（現、中央公論新社）の第30回女流新人賞を「答えて、トマス」（新風舎刊）にて受賞。その後執筆活動に入る。アメリカ人の夫とともに、ロンドン、チューリッヒ、東京などに住み、1996年より香港在住。香港大学在学中に季刊誌「陰陽」に戯曲「Takeru's Pinch (タケルのピンチ)」を発表。その後は定期的に、香港の文芸誌「Imprint」に英語作品を発表している。

ココ・リクター

アメリカ出身で、弁護士として活躍。現在は主に短編小説、戯曲を執筆。初の小説「テンプティング・ザ・ドラゴン」を2015年にインクストーン・ブックスより出版。香港の書店およびアマゾンにて販売中。香港の文芸誌「Imprint」および「香港ライターズ・サークル・アンソロジー」に短編を発表。香港大学にてＭＦＡを取得。在学中にリンゼイ美恵子と知り合う。香港ライターズ・サークル副会長主席。ウェブによる「ココのクイックリード」www.cocosquickreads.com 主催

Profile

Mieko Lindsay

Born in 1961 in Tokyo, Mieko graduated from Aoyamagakuin University and later obtained an MBA from Sheffield University in England. Mieko won a New-Woman-Writer's Award from Chuo-Koron Company in 1996 for her short novel titled *Kotaete Thomas* (Tell Me, Thomas). Recipient of an MFA in creative writing from Hong Kong University in 2011 and auther of *Takeru' Pinch* published in HKU's *Yuan Yang*. She has been contributing stories to *Imprint.* She also translates between English and Japanese.

Coco Richter

Originally from US, where she practicsed law. Coco writes short-length fiction and short screenplays. Her first novel, *Tempting the Dragon*, was published in 2015 from Inkstone Books and is available in Hong Kong bookstores locally and on Amazon. Her work has also appeared in *Imprint* and in *Anthologies of the Hong Kong Writers Circle.* She has an MFA from the University of Hong Kong where she met Mieko, and she is Vice-Chairman of the Hong Kong Writers Circle. Find her at http://www.cocosquickreads.com.

『だから（香港のアパートで）
犬は飼えないの！』
That's Why You Can't Have a Dog in Hong Kong:A Memoir

著　リンゼイ美恵子
By Mieko Lindsay

英語監修　ココ・リクター
English Supervising Editor　Coco Richter

2017年12月20日　第1刷発行

カバーデザイン　　Gai DESIGN　初谷 晴美

発行者　山田晋也
発行所　あるまじろ書房株式会社
　　　　本社住所　〒110-0005　東京都台東区上野2-12-18
　　　　　　　　　　　　　　　　池之端ヒロハイツ2階
　　　　編 集 部　〒942-1354　新潟県十日町市福島 864-4
　　　　代表電話　03-4405-7935　　FAX 03-6869-5308
　　　　URL　http://www.arumajiro.co
印　刷　中央精版印刷株式会社

落丁・乱丁本はお取り替えいたします。私的使用も含め、
本書のコピー、スキャン、デジタル化などの著作権法上での例外を除き、
無断複製を禁じます。
©Mieko Lindsay 2017　Printed in Japan
ISBN978-4-904387-10-8